MW00585767

Family Friendly Farming

Family Friendly Farming

A Multi-Generational Home-Based Business Testament

Joel Salatin

Polyface, Inc.
Swoope, Virginia

Editing and book design by Jeff Ishee

About the cover:

The Salatin Family in their front meadow at Polyface, Inc. in Swoope, Virginia. (l-r) Joel's mother Lucille, Joel, wife Teresa. (kneeling) children Rachel and Daniel.

Library of Congress Control Number: 2001130535

ISBN: 0-9638109-3-6

Other Books

by

Joel Salatin

Pastured Poultry Profit$
Salad Bar Beef
You Can Farm

All books available from:

Acres USA	1-800-355-5313
Chelsea Green	1-800-639-4099
Quit You Like Men	1-800-311-2263
Stockman Grass Farmer	1-800-748-9808

or by special order from your local book store.

Contents

Romancing the Next Generation

Family Development

A Sacred Work

Multi-Generational Transfer

Acknowledgments

I am most grateful to Teresa, my tender wife, who keeps me true to the ideas expressed in this book. Her consistency and nurturing toward our children have helped round off the sharp corners.

Our children, Daniel and Rachel, are truly the joy of our lives. When parents of adolescents shake their heads in dismay, I confess that I have not a clue what they're feeling. By the time this book goes to print, Daniel will hit 20 and Rachel will be 14. I don't deserve kids this wonderful, but we've had time together, and that has made all the difference.

My father William and mother Lucille shaped my character and created within me a desire for excellence. That standard includes marriage, children, business and environmental care.

To all who have led the homeschool movement, from Raymond and Dorothy Moore to John Holt, I extend deep appreciation. The Home School Legal Defense Association, the publication Growing Without Schooling and my mentor and friend, Rick Boyer have all encouraged me in thinking outside the box.

Whenever I quit farming for a few weeks and write a book, an extra burden falls on our apprentices. I thank Jason Rowberg, Seth Luerssen, and Chris Gowen for picking up the slack.

For my environmental thinking, I borrow heavily from those who articulate the arguments for living more lightly on the land. Sir Albert Howard, John Muir, Wendell Berry, Charles Walters Jr., Allan Nation, Louis Bromfield, Paul Hawken--the list could go on indefinitely. I borrow the best and disregard the rest.

To the many friends in our fellowship group and beyond whose conversations help shape my thinking, thank you. I appreciate our customers, without whom we would not have a viable farm enterprise. And finally, thank you Jeff Ishee, for the desktop publishing wizardry that turned this raw manuscript into a marketable book.

Foreword

It is no secret that family farms are in trouble. With the average age of U.S. farmers now over 60 years, the next few years will witness tremendous changes of land ownership in the countryside. For many of these lifelong stewards of the land, no wish is more deeply held than for a son or daughter to continue the family tradition. But this dream is often bittersweet, as few would advise their children to enter such a troubled industry.

The state of many American families is no less troubling. With both parents working full time, families increasingly struggle for meaning in life and values for their children . . . that is, between trips to the mall and running kids to soccer, dance, karate and baseball.

Joel Salatin and his family have demonstrated that there is an alternative to this desperate chase. They have shown that it is possible to raise healthy children who possess a zest for living, a natural curiosity, and love and respect for family and community.

And they do this while working alongside each other.

A visit to the Salatins' Polyface Farm, nestled in Virginia's lush Shenandoah Valley, is an antidote for cynicism and despair in this modern, harried world. Joel speaks frequently on his innovative systems of agricultural production. The methods he and his family have devised to raise food -- and make money -- are elegant in their simplicity and breathtaking in their sophistication. Through respecting nature, the Salatins allow each living creature to fulfill its God-chosen directive in life: Cattle graze, chickens scratch, hogs root, and rabbits, well, they multiply. And the land continues to regenerate and improve itself. Every aspect of the farm is safe and wholesome for children -- there are no poisons, no dangerous equipment, livestock confinement buildings, or any of the other sins against decency and humanity that so-called "modern" agribusiness have foisted on the countryside.

Joel's previous books have focused on these issues -- teaching other farmers how to raise livestock naturally, how to gain control of local markets, and how to actually earn a city-size income through farming. The book you are about to read springs forth from an area much closer to Joel's soul. This book does not focus on the "how" of farming, but the "why." For without our closest relationships in life, what is life? The Salatins actively practice what they preach. They invest in relationships, not things. Witness one of the farm's famous field days, in which several hundred pilgrims journey from across North America to the mecca of sustainable innovation. You'll see a few dozen neighbors, friends and church members cooking up a feast and having the time of their lives. And it is certainly no exaggeration to say that few parents could claim to be more proud than Joel and Teresa Salatin. Their children, Daniel and Rachel, are delightful, not as politicians-to-be, nor future Carnegie Hall performers or Olympic champions, but as human beings -- intelligent, innovative, curious, friendly, hard working, mature, laughing, loving, well-adjusted individuals who value God, their family and the community over fast cars and hollow dreams.

As many people ponder their own lives, their families, and the kinds of lessons they are teaching their children, the thought of an idyllic life in the country often surfaces, one that emphasizes truth and substance over hollowness. For many people, Joel's words will be life changing. But even if you don't choose to move to the country to raise a family, you'll find lessons here you can learn. And there is no better teacher than model father and farmer Joel Salatin.

Learning about the Salatins' methods of family-friendly farming gives us a sense of hope for the future of agriculture. And for many families seeking a more meaningful way of life and a better future for their children, it is a salvation. For this we all owe a very great debt of thanks to Joel Salatin and his family.

Fred C. Walters, editor, *Acres U.S.A.*

Acres U.S.A. is a monthly magazine of ecological agriculture. Fred is the second-generation manager of this family business.

Introduction

Would you like to wake up each morning and step into an office of rolling green pastures, verdant woodlands, and shimmering ponds? And then as you worked in this office, would you enjoy the company of your children and a loving spouse?

Western culture at the dawn of a new millennium yearns deeply for family ties within the worldwide web. The web represents globalism and everything techno-glitzy, but it moves, grows and grinds along without regard to Mom, Dad, Brother and Sister. For all that it promises to bring into the home, it cannot minister to the deepest needs of the human spirit.

The average American home is unoccupied for most of the day. And even when someone is home, other members of the family are scattered hither and yon. Internet connectedness is eclipsed by deep disconnections in marriages and familial relationships.

Couples routinely hit their mid-40s only to discover they

hardly know each other or the children they bore. With mid-life feeling like a storm-driven ship, these couples frantically seek a haven. The children cry for attention, but Mom and Dad are too career busy to give it. Families spin out of control and crash land, casualties of a society gravitating toward egocentricity and shallow values.

Take heart, though; all is not lost. Every movement creates its anti-movement. The lack of personal fulfillment in modern culture is illustrated by crime, abortion, divorce, retirement communities, and teen centers. These heighten awareness within the human soul, creating desire for alternatives.

A real alternative is the family friendly farm. The most successful juvenile delinquency treatment programs in the nation involve agrarian-based rehabilitation. Even without family, people in touch with plants and animals enjoy a soothing caress from life's hand. The Creator can reach gently into the human spirit through the eyes of a little chick or the magnificence of a fruit-laden grapevine.

Because my background is in farming, I draw from it for metaphors. But make no mistake, this book is geared toward any home-based multi-generational business. The principles apply to plumbers, advertising agencies and landscapers. Truth is truth.

Although I have written this book from a Judeo-Christian perspective, you will find that my Christian libertarian capitalist environmentalist bias will both endear me to you one moment and inflame you the next. All of us approach life from a certain bias and we may as well acknowledge it in humility. I only confess to being honest, eclectic, and still learning. My views tomorrow will differ a little from my views today, but not in the bedrock values regarding God, family, business, morality, and the environment.

When I wrote YOU CAN FARM three years ago, I thought

it would be the most personal treatise I would ever write. How wrong I was! I realize that in this new work I have opened myself up and made myself vulnerable for all the world to see. But it will all be worth it if, wherever you are in the cycle of life, you will be challenged afresh to strengthen the threads that tie together the fabric of a culture--family, business, environment.

What good are strong families if hog factories destroy the water? What good is saving the environment if children can't romp and play on the landscape knowing that Mom and Dad love them unconditionally--and that Mom and Dad are in love? And what good is being a millionaire when your 16-year-old son sneaks around behind your back and makes snide remarks? Many a successful business executive would trade it all to have well-adjusted, respectful, gung-ho youngsters in the home.

This book is about restoration and opportunity. Any time you feel like I'm pointing a finger at you, remember that three are pointing back at me. By putting it in writing, laying it all on the line, I've left a blueprint of my heart's desire for my children and grandchildren.

May God's richest blessings abide on your family. May your business serve your community in truth and integrity. May your farm radiate vitality, health and wholesomeness. May your children enjoy working with you. And in your golden years, may your children and grandchildren revere you, enjoying together a nurtured landscape and business ministry.

Joel Salatin
July 2001

Perspective

Chapter 1

My Vision

My vision contains three distinct themes. The first one has to do with multi-generational agrarian home-based business. The second deals with how best to populate the landscape with loving stewards. And the third concerns insuring a clean, nutritious food supply for our grandchildren. Family friendly farming offers a pathway to actualize all three themes.

Perhaps a short excursion into cultural history can help set the stage for the vision presentation. Throughout most of history, a person's identity could be found in a tribal group. To be sure, kingdoms have spanned many cultures and tribal groups throughout history. But by and large people did not move too far from their roots.

Even explorers claimed territory for their country, their local culture. When we look at the historical norm spanning human history, we see distinct groups of people with cultural and familial identity practicing trades multi-generationally. Non-industrialized cultures today still exhibit these characteristics:

1. Tight-knit family groups.
2. Small business.
3. Multi-generational business, trade or land control.
4. Elderly usefulness, reverence and care.

An anthropologist would probably have a longer list, but this is enough to get the picture. The industrial revolution and its offspring, the World Wide Web and globalization, fundamentally changed the loyalties, size and makeup of human identity. Perhaps the word identity is too strong, but what I'm getting at is that the daily fundamentals of life on which people put their attention are radically different. Consider a short list:

1. Geographically distant families
2. Multi-national corporations and the global economy
3. Job-jumping and vocational short-timing
4. Elderly obsolescence, retirement, and disrespect

If you'll look at both lists, you'll see that a fundamental cultural change has taken place. It has happened a little bit at a time, like the frog in the slowly heating water, but looking back we can see a definite shift.

Another way to see the shift is to just think back about home life in early America compared to today. Mom ran the home; Dad ran the cottage industry; children did chores, studies and played. You could visit the average home at midday and be greeted by people.

Today the average home at midday is vacant. Mom and Dad are at work and the children are at school. When they converge on the home, they are seldom all there together and home is viewed as a pit stop between races. The race may be the expressway on the way to the computer cubicle or it could be the race to the soccer game. The children are seldom interested in what Mom and Dad do: that's considered archaic.

The home's reason for existence is virtually nonexistent. A hotel room would work as well. The interests of the family, even if it is still a nuclear family, do not center around home and hearth, but are all out there somewhere beyond the front door. And if that weren't enough, we bring the world in on the web. Rather than talking or reading aloud, the family's evening, even if it is spent under one roof, consists of television in two rooms and the web in the third room with the video game in the other room.

The bottom line is that the goal of our 21st century technological culture is to get a good enough job working for someone or something wealthy enough to live in a school district good enough to give our kids an education good enough to get a job good enough 1,000 miles away from home so that they can afford to put us in a nursing home when we cease to be productive enough. As a result, middle-aged adults resent the elderly burden; the young have no real link to their grandparents except through holiday festivities; parents don't know their children and vice versa; and marriages blow up in this societal pressure cooker.

My vision is that the grandchildren will play around my legs and have to draw straws to see who gets to spend the day with grandma and grandpa. And our children will revere and care for us as we age, soliciting our advice and leveraging our experience with their youthful energy. These children, spending all their time with parents and grandparents, will become, in the words of Wes Jackson, "native to this place," developing an ethic for the local ecology based on multi-generational information. And neighbors will flock to our farm for the best food available, giving their children the gift of nutrition and health.

One of the habits Stephen Covey articulates in his seminal book Seven Habits of Highly Effective People is that we must start with the end in view.

If we want to be revered by our middle-aged children; if we want to spend copious amounts of time with our grandchildren; if we want to impact our community with a golden rule business; if we want our business to continue beyond our lifetime; and if we want to populate rural America with land loving stewards, then we must institute principles now to create that end. It will not happen by accident. A certain protocol, a certain procedure, can create an environment conducive to such an end. And it is all the nuances of that protocol that this book addresses, or at least all the ones I can think of right now.

Our cultural paradigm offers electronic connections, but never have families been more disjointed. As a culture, we seem to be on an accelerating train away from the things that really matter. Deep personal and communal relationships that form societal glue, that form reverential bonds to our local environment, can scarcely survive the globalist's agenda. Global gatherings addressing environmental, societal, moral or cultural concerns offer bureaucratic, top-down solutions that never work because, even if they started sincerely, they become bastardized as they filter through the powerful global mind set.

Does anyone believe family ties are stronger now after worldwide family-oriented conflabs? How about desertification and chemical usage? How about adulteration of the food supply via genetic engineering or food-borne bacterial contamination? How about an agricultural land ethic? Even the most avid Nature

Conservancy member realizes that the two percent land area mapped into the critical habitat protection plan is completely dependent on the surrounding 98 percent of farm, industrial and urban land.

Perhaps you're wondering at this point where the farming comes in. After all, we've scarcely talked about farming. Now it's time to acquaint you with my personal bias: family friendly farming offers a potentially effective model for addressing all these issues. But something is desperately out of whack when the average age of the American farmer is nearly 60 years old and trending upward.

Business books agree that anytime the average age of the practitioners in a sector of the economy rises above 35, that sector is in decline. Vibrant economic sectors are dominated by mid-life people who have both the experience and the physical energy to be highly creative and productive. Looked at another way, only when older experience is leveraged through youthful energy can true production occur. Attend an average meeting of cattlemen and you'll be hard pressed to find anyone younger than 40.

This is a recent phenomenon. The virile, muscled body of the American farmer that dominates pictures up until the 1960s has been replaced by aging, tired, overweight codgers. Visit any livestock sale barn in the country and you'll see this stereotype living before your eyes. This change, just like the de-emphasizing of the home, has occurred slowly but systematically.

The agrarian culture that Thomas Jefferson envisioned and which dominates all cultures in their early years has become a liability at the turn of the millennium. Perhaps nothing is more endangered than the American farmer, who doesn't even merit counting on the census anymore because so few exist. We count homeless, single mothers with dependent children, and nursing home occupants, but not farmers.

Farming is now the premier example of a vocation that cannot and does not reproduce itself. No other profession compares to it in non-generational succession. The number of elderly farmers not being replaced with any children is epidemic. The number of farmers has dropped precipitously in the last few decades. When Ed Faulkner wrote Plowman's Folly in 1943 it sold half a million copies the first year and had to be reprinted 5 times. Can you imagine a book by that title selling that many today? It would not even be published. Every full-time farmer in the country would have to buy it in order for it to sell that many copies.

Replacing farmers with machines and centralized factory farms has depopulated the countryside. In many rural areas across the grain belt, school districts have amalgamated three into one in order to have enough kids to fill one school. Routinely farms that at one time supported 10 families now support only one.

I ran across a gem recently that helps explain where we are. Written by three professors at Cornell University, the 1942 agriculture textbook Farm Management and Marketing opens with chapter 1 answering the question: "What is Farming?" It spends the bulk of the chapter differentiating farming from industry. The nutshell of the thinking was capsulated in a list of four elements. Here is the seminal passage:

"*Farming is not adapted to large-scale operations because of the following reasons:*

1. Farming is concerned with plants and animals that live, grow and die.

2. Farming is greatly dependent upon climate and weather conditions.

3. Farming requires the making of many quick decisions by the individual worker.
4. Farming requires enough personal interest on the part of the individual worker so that he will assume major responsibilities and work harder or longer hours on some days than on others."

We've come a long way, baby. Can you imagine that message being preached in the modern agriculture classroom? This issue brings us now to the second basic theme addressed at the beginning of the chapter. Remember the first was the viability of the multi-generational small farm. The second is the most sustainable model for populating the countryside with land-loving stewards.

The issue here is not just what you may or may not want to do with your family. It is much bigger than how an individual family desires to live, grow old and leave legacies. It is more important than doing something with your kids and your spouse that will unite you as a family. It is about how the American landscape will be loved and nurtured.

The capacity to care, to have a deep and abiding love for a place, is far greater in a person than a machine. A farmer can surely love a field far more than his turbo-charged diesel tractor. Even in a perfect world, the pre-Edenic world, God placed a person, Adam, to tend and dress it. A landscape devoid of people is not a properly stewarded landscape.

I know this puts me crossways of many environmentalists, but the idea that things left to themselves are far more perfect is utter nonsense. People have been touching the landscape since the beginning of time. Just to show how strong this urge is, consider that even in today's designated wilderness areas, we practice fire suppression. To be really natural, we should let those fires burn as big and as long as they want.

When people try to pigeonhole me, I simply respond: "I'm a Christian libertarian capitalist environmentalist." The designation, I hope, shows that I do not cotton to single-issue solutions like buffalo commons for the prairie and old-growth worship for the arboreal forests. By the same token, I do not believe the current logging practices have been environmentally friendly nor that plowing up the prairie was the right thing.

The cultural pendulum, as well as my personal pendulum, seldom hangs straight down. It's usually over to one side. We all must be big enough to appreciate the fact that our pendulums are seldom hanging perfectly balanced. We're either too hard or too light on the children. We're either working too hard or being lazy. Culturally, we do the same thing. One of my favorite examples of this is Upton Sinclair's 1905 blockbuster The Jungle, exposing the filth and animal abuse in the Chicago slaughterhouses. Within a few months of the book's publication, beef purchases nationwide dropped more than 30 percent.

In 1906 the first meat and poultry inspection laws were passed, supposedly to guard against these kinds of abuses in the future. Of course, the regulations were added to through the years until now a person can't even perform one of history's most premier rural rituals--the fall hog killin'. Oh, the hogs can be dressed alright, but nothing can be sold without a $1 million, quintuple-permitted, properly zoned, monolithic factory. Does anyone believe that a backyard-dressed pig in a community or familial hog killin' is dirtier or more unsanitary than what comes out of extruded, amalgamated, irradiated, chlorinated, adulterated fecal factories overseen by bureaucrats? Is a bureaucrat more honest than a farmer? Of course not.

But instead of society waiting for the marketplace to work on the Chicago slaughterhouses and the abuses in the meat packing industries, letting private watchdog groups put seals of approval on extraordinarily good practitioners, society knee-jerked into a bureaucracy that today is the single largest impediment to getting clean meat, poultry and milk into the marketplace.

Sometimes the pendulum shift is not quite as obvious. Every paradigm creates its anti-paradigm. A century or so ago everyone ate locally produced, unprocessed food. Laura Ingalls Wilder's Little House on the Prairie books introduced me to the word "boughten." Remember how she was enamored of boughten white sugar and boughten flour as a substitute for the drudgery of maple sugar and coarse whole wheat flour? That trend climaxed in the 1950s and early 1960s with the notion that nothing could be as barbaric as breast-feeding babies.

We raised a generation on formula and processed baby food. But then came Vietnam, flower power, peace symbols and hippies. Rachel Carson's Silent Spring hit like a bombshell and the environmental, back-to-the-land movement was born. The embryonic ideas kindled at that time are bearing fruit today in the trend toward green everything, wellness centers, healthy food, recycling, and urban flight.

It is fascinating that in 1800 only about 5,000 people owned all the land in England. In 1900 it was owned by more than 1 million. Things do change. The pendulum does swing. I think it is a mark of wisdom and maturity to enjoy the swing and not demand immediate action. I get crucified when I refuse to agree that our rivers would be as clean without the Environmental Protection Agency. The fact is that I grew up shooting sparrows with a .22 rifle. Our kids wouldn't think of doing such a thing. Are sparrows sacred? No, but in the 1960s and 1970s we didn't have wildlife refuge centers and school assemblies devoted to protecting the environment.

It took a degree of pollution to spawn the environmental movement. The EPA was simply an unnecessary and now more often than not an impediment to cleaner water and air. Multinational corporations now have sustainable policy projects, in-house groups watch-dogging and writing protocol to maintain eco-friendliness within the company. Does this mean all companies are lily-white? Is no dumping occurring? Of course not. But that is why the free flow of information, a free press, and an unharried, unhurried populace with time to self-educate is necessary to continue moving forward.

So as the information age replaces the industrial age, at least in the US, and where your work is no longer important as long as you're linked up, the depopulation of the countryside that has occurred during industrialization will be replaced by a repopulation. This repopulation is coming from two sectors. One is the highly linked folks who have their laptops and will travel. The office is anywhere you can hook up the laptop to the web. It can be on the beach in the Bahamas or a mountain hideaway in Appalachia.

The second group is comprised of folks who have tasted and seen the inside of a Dilbert cubicle at the end of an expressway and are looking for something more out of life. Many are mid-lifers whose 20 years working for the company store has been touched by outsourcing, downsizing or restructuring. Add to that the fact that when they come home their early teen children are wiggling to a

Walkman and in the privacy of their souls they ponder what kind of grandchildren they'll have.

Remember, every paradigm creates its anti-paradigm. The depopulation of the countryside absolutely will be reversed. What this means is that the family unfriendly paradigm of the past: the model that has sent the children to the cities, that has created an elderly farmer, that has resulted in countless estate sales and the centralizing of agriculture, is ready to be replaced by a family friendly model.

And what about farming? Why farming more than plumbing, wiring or masonry? Nothing especially, except that farming is what I'm familiar with and historically, even tradesmen kept a cow and planted a garden. This compartmentalized notion that a culture can be healthy when less than 2 percent of its population dabbles in the life-giving art of food production is crazy. To think that people can be as out of touch with their living world as most people are today and still revere life, each other, and the plants and animals that sustain us is lunacy.

Where is the greatest dysfunction in our culture? Not in the countryside. It's in the city. I'm not suggesting that everyone should move out of the city, but I am suggesting that cities of the magnitude we've created are not healthy for our world or our culture. And they certainly aren't healthy for our families.

The mystical haven mantra that surrounds the pastoral life is shared by more people than you can imagine. What cuts it short is the stark notion that a farm life is full of unrewarded toil, low pay, dirt, dust and drudgery. But this is not necessary. And the solution is not biotechnology, clones and slaves to the vertically integrated multi-national corporation.

And this brings us to the quality of our food supply issue, the third theme in this vision. How will our grandchildren be assured real food? How will they be protected from the food-borne bacteria, mad cow disease and the host of other new living toxins in our food supply?

In 1996 the Government Accounting Office released a report on food safety dealing with food borne illnesses, showing that the risk is increasing. The report cited several factors for this increase:

> *"First, the food supply is changing in ways that can promote food borne illnesses. For example, as a result of modern animal husbandry techniques, such as crowding a large number of animals together, the pathogens that can cause food borne illnesses in humans can spread throughout the herd.*
>
> *Because of broad distribution, contaminated products can reach individuals in more locations. Mishandling of food can also lead to contamination.*
>
> *For example, leaving perishable foods at room temperature increases the likelihood of bacterial growth, and improper preparation, such as under-cooking, reduces the likelihood that bacteria will be killed and can further increase the risk of illness. There are no comprehensive data to explain at what point pathogens are introduced into foods. Knowledgeable experts believe that although illnesses and deaths often result after improper handling and preparation, the pathogens were, in many cases, already present at the processing stage. Furthermore,*

the pathogens found in meat and poultry products may have arrived on the live animals.

Second, because of demographic changes, more people are at greater risk of contracting a food borne illness. Certain populations are at greater risk for these illnesses: people with suppressed immune systems, children, and the elderly. In addition, children are more at risk because group settings, such as day care centers, increase the likelihood of person-to-person transmission of pathogens. The number of children in these settings is increasing, as is the number in other high-risk groups, according to CDC [Centers for Disease Control].

Third, three of the four pathogens CDC considers the most important were unrecognized as causes of food borne illness 20 years ago-- *Campylobacter, Listeria, and E. coli* *0157:H7.*

Fourth, bacteria already recognized as sources of food borne illnesses have found new modes of transmission. While many illnesses from E. coli 0157:H7 occur from eating insufficiently cooked hamburger, these bacteria have also been found more recently in other foods, such as salami, raw milk, apple cider, and lettuce. Other bacteria associated with contaminated meat and poultry, such as Salmonella, have also been found in foods that the public does not usually consider to be a potential source of illness, such as ice cream, tomatoes, melons, alfalfa sprouts, and orange juice.

Fifth, some pathogens are far more resistant than expected to long-standing food-processing and storage techniques previously believed to provide some protection against the growth of bacteria. For example, some bacterial pathogens, such as Yersinia and Listeria, can continue to grow in food under refrigeration.

Finally, according to CDC officials, virulent strains of well-known bacteria have continued to emerge. For example, one such pathogen, E. coli 0104:H21, is another potentially deadly strain of E. coli. In 1994, CDC found this new strain in milk from a Montana dairy."

Doesn't all this make you feel safe? Since this report was released, the food industry has become far more centralized, with mergers and buy-outs happening every day. When Cornell showed that forage-based beef had practically no E. coli you'd think the feedlots would have jumped for joy and instituted a new forage-based protocol. No way. They pooh-poohed the report and said such preventive measures would be impractical. Instead of decentralizing and bringing clean animals and plants to the marketplace, the industry is on a fast pace toward irradiation, which is euphemistically being labeled "cold pasteurization."

If bread is the staff of life, then fundamental to a quality of life is eating good food. Something as basic as good food is now in very real jeopardy. Oh, we have mountains of stuff coming out of Archer Daniels Midland and other collaborators in the devil's pantry, but its nutritional and safety integrity could not be more questionable. What many people eat as food today will not even rot on the shelf. You can leave it sitting on a kitchen counter and it does not even contain enough life to decompose. Will that give us healthy grandchildren? I think not.

And so as these themes of multi-generational agrarian home-based business, ecology and clean food converge, the family friendly farm rises to offer a vision for our times. Perhaps never before have more pieces been in place for a family friendly farm.

A balanced approach toward the use of technology, including the web, the use of machinery, and the integration of the farm with its community of resources, animals, plants and urban refugees can yield a rewarding multi-generational agrarian business hardly imaginable throughout most of history. The environmental and food awareness movements in our culture, the technology and biological knowledge we now possess, and the equity available in our economy are all intersecting in wonderful ways to create plentiful ingredients for the family friendly farm.

My Vision

Chapter 2

Growing Up on Our Farm

As far as I can tell, my farming lineage is unbroken to as far back as when almost everyone farmed. But it hasn't always been family friendly. My great-grandfather was killed blowing out tree stumps from a field he was clearing.

My grandfather tried to make a go of farming after his father-in-law (my great-grandfather) was killed. But he couldn't make it go, so he moved to town on a couple of acres where he kept a half-acre small fruits and vegetable garden. A chicken house in the corner acted as a garden garbage disposal and several hives of bees produced honey.

My father and mother headed to Venezuela, South America to farm and after a decade got caught in the 1959 revolution and had to flee. Being ugly American capitalists, our family was a prime target for the left-wing revolutionaries. I was four when we returned to the States and left that chapter behind. Always looking for the new thing, Dad viewed the developing country as a truly free marketing opportunity. But political instability dashed those dreams.

My brother Art, who was three years older than I, and my sister Loretta, five years younger, grew up here in Virginia's Shenandoah Valley after we returned stateside. I was four when we looked at farms in a wide arc from North Carolina to Pennsylvania--Dad wanted to be a day's drive from Washington, D.C. because he was meeting with Venezuelan and U.S. government officials trying to get a settlement for the land taken from us there.

We found this worn-out farm in July 1961 and moved in to start over. I well remember Dad seeking advice from people in the community. Since I was not in school, I accompanied Dad and these experts in our walks over the farm and listened to their advice. The advice was universally in favor of things Dad opposed: graze the forest, plow and plant corn, build silos, build feedlots, etc.

Absentee owned for nearly 50 years and then owned by non-farmer rich folks in a quick succession of three four-year stints beginning in 1950, the farm had huge gullies and rock outcroppings. Although this made a great place to hunt for fossils among the rocks and play cowboys and Indians in the gullies, it was not conducive to profitable production. We bought some Hereford cows right away and began. One fence separated roughly one-third of the pasture from the other two-thirds.

An old Case hay baler powered by a two-banger air-cooled Wisconsin engine, a retrofitted-to-tractor horse-drawn hay rake, a three-point hitch cycle bar mower and 1952 model gray and red Ford 8N rounded out the machinery inventory. We bought the equipment from the previous owner.

The first machine Dad bought was a 1952 one-and-a-half ton International dump truck. One of my first and most pleasant memories is, as a four-year-old, driving that truck while Dad threw off hay to the cows during that first winter. I wasn't tall enough to press on the floor pedals, so Dad put it in granny gear (it was four-speed with two-speed rear axle), set the throttle to keep the engine running steady, then jumped onto the running board with the door open while I stood on the seat and steered. Then he would jump into the bed and throw the hay off the back in a long windrow. When he finished, he climbed back onto the running board, I stepped over to the other side of the seat, and he stopped the truck.

I remember feeling unbelievably important doing this job. One morning as he was walking back through the cows checking them, he noticed a fox standing over by a grove of pine trees. We eased over there and the fox just sat there, completely motionless. Dad picked up a stout baseball bat sized tree branch and told me to stay put. I'll never forget him walking up to the fox and clubbing it over the head with that stick. The fox never moved a muscle, obviously rabid.

Why belabor this story? Because it brings out the very best in family friendly farming. As a little chumper, I felt important by doing a man's job--driving the truck. I watched Dad, the livestock and family protector, deal with a potentially life threatening situation with courage and dispatch. I saw the realities of life, death, and proper stewardship. The fact that I can remember this as if it were yesterday shows just how powerful real life experiences are.

I'm not suggesting that these principles of a child's self worth, a father's protection and courage, and nature's vagaries cannot be taught in a city. But the visceral, sensual experience adds a power to it that makes growing up a real experience, not a virtual experience where even taking out the garbage is viewed as too much work.

A recent news program carried the findings of a study tracking the life of the 1950s Whiz Kids, those brainy youngsters who wowed the nation in one of the first knowledge-based TV shows. Most shows since have featured adults, except for the youth portion of Jeopardy. Anyway, what these researchers found was that these brilliant youths had the same success and failure rate as less smart kids. A couple died after becoming alcoholic bums and a couple distinguished themselves in their various fields of expertise, one even receiving a Nobel Prize.

The study concluded that life's success has very little to do with smarts; it has to do with two things: perseverence and work. Amazing. It reminds me of the fact that in our techno-crazed world parents love to extol their children's smarts, but they aren't as interested in their virtue. Academics do not equal character. And while other kids may have been stacking alphabet blocks in kindergarten that day, I was participating in the great principles of life, creating memories that would last a lifetime. And not getting teased about buck teeth or big ears.

The study also found that the single biggest ingredient for creativity was not IQ, but humor. Imagine that. Humor. It's such a surprise ending to the study that it's almost laughable. And that brings me to a couple more stories.

After that first year Dad realized that until the mortgage was paid off, the farm could not earn a salary--at least with his knowledge level. He and Mom both worked off the farm and paid it off in eight years. Mom taught high school health and physical education; Dad did accounting and bookkeeping for a couple different accounting firms. We needed a second car so he bought an old 1957 Plymouth from the neighbor for $50.

Ever the frugal accountant, Dad didn't want to have a pickup truck in addition to the two cars. But we needed a vehicle to haul baby calves, hardware, chicken crates and whatever else a farm needs to haul. Creating the ultimate multiple use machine, he ripped off all four doors and took out the seats. This created one big inner room that was about the size of a pickup truck bed.

He sat on a galvanized metal bucket and we kids would ride in old apple crates. This was in the early 1960s before mandatory seatbelt laws and automobile inspections--back in the good old days before government intruded into every facet of our lives. It is the freedom to do the unconventional that creates an entrepreneur-friendly commercial environment. Anyway, he used this car for both driving to work and the farm stuff.

He and Mom were both partial to bow ties rather than neckties as professional attire. Imagine the picture of dad in top hat and bow tie, white shirt and suit, tooling around town to his different accounting clients in this ragamuffin door-less, seat-less car, sitting on a beat up galvanized metal bucket. Art, my brother, and I played trombone and trumpet respectively in the school band and would have to stay after school from time to time to practice for upcoming concerts or music competitions.

We'd wait out by the road in front of the school for Dad to pick us up on his way home from work. Duly told about the pickup that morning before heading to town, he would normally forget about it with the press of things on his mind by the time he headed home. As he went sailing past us in this car, we'd wave and yell to flag him down. It always took about 100 yards for him to realize that those two lunatics he just passed were his own sons, and as this thought registered he'd mash the brakes and the car trunk would fly up in one grand jalopy show. We'd jump in the apple crate and off we'd race toward home.

That car holds fond memories for me. When it finally graduated to full-time farm vehicle, we boys got to drive it, racing the pony or doing wheelies in the pasture. If you think these lessons are lost on the young, think again. When I got out of college and bought my first car, it was a $50 1965 Dodge Dart from a neighbor. The apple didn't fall far from the tree. Teresa and I took our honeymoon in that old car; kept it three years and sold it for parts after it threw a rod--for $75. How about that?

When Dad finally decided we needed an acetylene torch in addition to the arc welder, he bought one and hooked it all up. For ten years, whenever we cut steel in the shop it was either with a hand hacksaw or with the welder turned up high. That gives a crude job, to be sure, but it sure beats a hacksaw if pretty isn't necessary.

Anyway, I'll never forget him lighting the torch, adjusting the flame, then cutting a neat slice through a piece of quarter inch flatbar. He finished in a couple of seconds, tipped the goggles up on his forehead, and grinned with the mischievousness of a three-year-old finally reaching the cookie jar: "Now that's power, man. That's power." Humor. It takes the pressure off your failures and keeps you under control during your successes.

The old Case hay baler we started with became more and more erratic. No amount of adjusting would produce a consistent tie and Dad began fishing for an efficient alternative. He bought a hay loader for $50 (are you beginning to see a pattern here?) from a neighbor and we began making hay loose. It certainly is a much simpler machine, and in the early years proved efficient enough because we simply didn't have much hay to make. Soil fertility left a lot to be desired. I remember as a 15-year-old mowing hay and losing my place because the grass was so thin. I'd have to sight on a grass clump down the field to know I was steering on the right mowing line.

We gradually became more efficient by laying a chain on the wagon bed before we started the load. When it was full, we would back into the barn and hook a cable to the front of the chain and pull the whole load off in one big jelly roll. One day we were plodding along in the meadow next to the public road. Dad was driving the tractor and I was forking the hay around on the wagon, building the load. Of course, these loads were always picturesque, looking like a bread loaf with shaggy sides. They looked like something right out of the 19th century.

Anyway, the neighbor went by with his newfangled round baler and big green John Deere. Dad twisted around in the seat and caught my attention to proudly proclaim, pointing first to the round baler and then to our hay loader: "Old! New!" What a hoot. That was his zest for the unconventional. Back to the future.

One of the most profound memories I cherish from late boyhood was a reward for picking blackberries. Our farm is 100 acres open land and 450 in woodland. When we came, it was 160 and 390, but we planted trees on about 60 acres to stop erosion. Hillsides had 12-16 foot deep gullies, creating a corrugated roofing effect. Bare, galded hillsides. Since horses didn't fall over on steep terrain, for a century farmers plowed these steep hillsides to produce the grains that gave the Shenandoah Valley its Civil War legacy

"Breadbasket of the Confederacy."

Many early spring days would find Dad and us boys with pine seedlings and dibbles, planting trees on these hillsides. Today many of these trees are nearly 12 inches in diameter, tall and straight. The ground underneath is a 6-inch deep carpet of brown needles. The erosion has long since stopped and we're in the soil reproduction stage, but the scars of land abuse will be there for centuries. Wearing out land is not family friendly.

This particular summer we had a bumper crop of blackberries and Dad told Art and me that if we picked so many gallons he'd do something special for us. I asked him years later if he had something in mind when he promised the reward, and he confessed that he had not a clue what to do if we came through.

We jumped on the challenge and picked and picked until our fingers were purple and swollen from stickers. But we made the volume and Dad had to make his promise good. All that forested woodland was completely inaccessible. It lies on the east face of Little North Mountain, which is 1000 feet higher than our house. Our farm is roughly two miles long and half a mile wide. A highway and railroad run at the foot of the other side of the mountain, which is much steeper than the eastern side on which our land lies.

Even though we had lived here for ten years, none of us had been up to see the acreage at the top of the mountain. Our familiarity stopped with the 200 acres of open land, eroded hillsides and adjacent woodlots. The other 350 acres were completely unknown and unexplored.

The reward Dad finally settled on, and we relished, was to

take us boys on a daylong hike over the mountain, beginning at the road on the far side, and ending here at the house. Mom drove us over and dropped us off at the edge of the road and we began the ascent to the summit. Dad had a survey map and compass to help us stay on course. We explored hollows and rocky outcroppings, looked at lots of trees and tried to figure out where we were on the map.

We eventually found our way home, tired but exhilarated. A few years after that, when I expressed to Dad that I wanted to continue farming as a life's vocation, he and I would try to walk the mountain sometime between Christmas and early January as an annual ritual. Over the years, we, as any farm family does, developed our own designations for different places. The names accumulated from different experiences and realizations.

Shangri-La was a long, well-watered draw that had especially large native white pines on a southwest slope. Dad always wanted to put a large pond in there, and I still do. The field by Jim was obviously a field adjoining our neighbor to the north; the field by Thompson adjoined our neighbor to the south. The flat field is a long, flat field behind the barn and the far flat field is a similar field, but farther away from the house. The squirrel tree is one of those quintessential landmarks, a massive white oak always covered with fat gray squirrels. Located at the bottom of a hollow where the terrain flattens into a fertile flat drainage area, it denotes a confluence of field, water and forest. About the first year we were here, we were making hay in a field and got caught in a terrible hailstorm. From then on that was the hail field.

Through the years, the entire farm became mapped in our memories, and we can describe any place, almost to the foot, by terrain, experience, or vegetation. This sense of place, this connectedness to a location, permeates my very soul. I believe something happens in the heart of a young person when youth turns into adulthood in an area where he is surrounded with parental memories. We live in an incredibly disconnected society, where location is as transient as the next job opportunity.

When Dad became ill we walked together and picked out the spot where he wanted to be buried. He and I went before the local government officials to plot off a family cemetery that would have enough spots for at least three generations. My tie here is practically indescribable. I can stop by Dad's grave any time. And there's not a place, not a spot, I can go on the entire farm without a memory of having been there with Dad.

Talk about roots. I believe the phenomenon of searching for family history and family trees in our modern culture has been created by the transience of place and family spawned in our industrial culture. Even when people stay close to parents, development alters the landscape to such an extent that in an urban setting nothing looks the same after thirty years. Bulldozers raze old buildings to make room for new ones.

But on a farm, where the landscape changes slowly and the seasonal cycles run true to course, the sight, smell, touch of the memory lingers for generations. Such powerful memories, such emotional ties, create a reverence for the preceding generation that mellows love affairs with fads and fancies. A value system can reach across a generation because the tie is more than an academic bloodline relation, sketched on a sheet of paper.

Perhaps this is why God gave Israel a clearly defined land area and then went on, within that area, to lay out a protocol for land use. Piles of stones clearly defined familial property lines. Every

50th year the property, which had been bought and sold perhaps a couple of times during that period, reverted to these ancestral families so that it could never ever pass out of the original stewards. The American pioneers, unfortunately, did not cultivate this sense. With new land always waiting out there on the horizon, American farmers treated the resource as something to use up and leave.

As a culture, we're still wrestling with this dominating, abusive heritage. Part of the push in both the 1970 and 2000 back-to-the-land movements, I'm convinced, was a cultural atonement, or cleansing, over the guilt of the rape that occurred during the first 200 years in this country. The notion among the colonialists that if a person healed with herbs, traveled by foot or canoe and kept his history in oral tradition rather than written tradition he was an uncivilized barbarian without any rights and due no respect is reprehensible.

For the pendulum to swing the other way, of course, required a degree of rape and pillage on the landscape, and a degree of disrespect for life, that awakened a new attitude among many people. We as Americans had to reach a collective stomach-turning point before land stewardship and the clean movement could take hold. The pendulum is swinging; only time will tell how far.

My childhood memories are replete with mental pictures of doing things as a family, whether shucking corn, dressing chickens or hauling firewood. We examined problems collectively, played collectively and worshiped collectively.

I can never remember having a sense that anything was for today only. The discoveries we made, the hillsides we reforested, the ponds we built and the compost we spread always contained a long-term thrust. I always had the feeling that today's work was not for us, for the moment, but was one more brick in the wall that would gradually take shape 200 years from now. Our fast paced lifestyle in the techno-full age is defined by the short-term. Computers have a shelf life of only a few months. Planned obsolescence is the name of the game. Development comes so fast we dare not invest in anything that doesn't pay back within a year or two.

Perhaps this attitude defines the frontier edge of any culture, but I think it damages a value system predicated on long-term sustainability. Why does discovering something new inherently jeopardize the value of the old? On our farm, we buy a new tractor with a front end loader to handle more compost and spread it more efficiently. That is using the new to respect the old.

Electricity on our family farm is not used to run fans and feed augers as it is used on concentration camp chicken factories. Instead, electricity is used to power special fencing to protect pastured chickens from predators and give the birds even a better life than they would have had in pioneer days. When the temptations come, and they always do, to run pell mell toward the newest gadget, glitz and sizzle, Dad's presence in experience and memory hovers over me like a wonderful protective hedge. It provides boundaries to thought and action that appropriately slows down reckless abandon.

Our family has no desire to buy other farms, build empires and gain market share. I can go back and review goals written before Dad died that identified additional resource synergy on this place. We have not yet begun to tap the opportunities afforded on this little farm. That will take another couple of lifetimes, and by then new things will develop that we haven't begun to think of yet.

Every child enters adulthood with special memories. I think that too often those memories come mainly from holidays, vacations or special events. What I've tried to portray in this chapter is that my powerful emotional memories have nothing to do with special times, but are instead a recital of mundane tasks done with long-term emphasis. Because each task held a philosophical underpinning for tomorrow, its significance carried an elevated value. And so instead of my most precious memories being relegated to festivals, holidays and vacations, they are a stream-of-life collection from the house I live in, the paths I walk and the places I see every single day. And that is rich.

I covet the same treasure chest of experience, memory and emotion for every adult. Most children on modern techno-farms want absolutely nothing to do with the farm. They want to get away as soon as possible. That is a tragedy, and a mind set that indicates something was significantly out of whack in that farming experience. I hope these stories and gut-level feelings show that a childhood on a family friendly farm can be richly rewarding. Romancing, even. And that is what this book is about.

Growing Up on Our Farm

Chapter 3

Money: Cultivating a Proper View

Family friendly farming is about more than money. Business is about more than money. But if we don't make a profit, all the noble goals in the world won't pay the bills. Cultivating a proper view of money, and its first cousin, material possessions, is critical in a multi-generation family business.

Who wants to be rich? I feel sorry for you if you want to be rich. Dad would tell us about one of his aunts, who at prayer time would always plead for the rich people who had to invest such time and energy worrying about their money. How to keep it. How to spend it. She truly felt sorry for them.

Jesus' admonition that "it is easier for a camel to go through the eye of a needle, than a rich man to enter into the kingdom of God" should give us all pause. Fortunately, He added the punch line: "With men this is impossible; but with God all things are possible." In any event, this admonition should give us all a healthy respect toward the importance we place on money.

Wealth does strange things to people, just like power and fame. Only the most uncommon people can handle any of these graciously. I've been privileged to know some extremely wealthy people. I find that as a rule they are no more or less gracious then people in all other spectrums of society. Studies like the book The Millionaire Next Door reveal that most wealthy folks did not aspire to wealth. Bill Gates, founder of Microsoft, certainly did not aspire to wealth.

Why would anyone cheapen their life by aspiring to wealth? A rich person who spends all his time trying to acquire more is no different than a poor person who tries to do the same thing, or who would act the same way if he had riches. Money is nothing more than a tool of exchange. I'd be happy if it didn't exist, but that's not going to happen any time soon. The challenge is to develop a healthy attitude toward it.

"I want, I want, I want" is one of the first phrases all parents deal with when their toddlers begin to talk. If it's not dealt with there, it will become a lifelong handicap. And here comes the pendulum again, trying to balance. We don't want to crush the spirit in the process of disciplining the desire. We want to preserve a passion for reaching for the stars without it being a me-sided property grab.

Motivational speaker Zig Ziglar says you can get where you want to go if you make sure enough other people get where they want to be. This brings up the servant's heart. I realize this can be construed as self-serving, but there is a fine line between selfishly getting ahead and unselfishly achieving goals.

Perhaps some very personal examples can help shed some

light on what I consider a proper attitude about possessions. This is a difficult discussion, because money is something we must use, and use wisely, but it's easily abused. And it's a hard master.

We're all familiar with the ditty about "keeping up with the Jones's." Many farmers buy things because that is what is expected from people in their socio-economic climate. One-tonner dually pickups have drained many a small farmers's cash reserves when a more conservative truck would do. Or subcontracting the hauling done. Or borrowing a friend's truck and bartering some milk or eggs to sweeten the deal.

The most vicious trap any of us can fall into, and farmers seem to be more prone to it than most, is this desire to have what the neighbor has. A corollary is the desire to be first. The first calf, the first corn, the first hay. Every one of these is completely uneconomical. I know a farmer who routinely wins the corn production contest. He identifies the couple of acres he plans to enter, and then babies it, overfertilizes, oversprays, and coddles that crop into first place. But he's quick to confess that those couple of contest acres lose him lots of money. It's just the notoriety of winning the contest that he's after. It has nothing to do with economical production.

The way these farm boys fawn over their pickup trucks is indeed something to behold. We farmed for 20 years before we even had a licensed pickup. But these young bucks look at chrome, cellular phones, stereo systems and paint jobs as if that's the secret to hauling firewood. All that gingerbread doesn't make the machine one whit more functional; it only makes the purchase price and upkeep higher.

Most of the neighbors around here own John Deere tractors.

I have nothing against John Deere, except that they seem kind of pricey to me. Anyway, I'm known in the community as the farmer that has the wrong color tractors.

My brother astutely noted one time, when lamenting that we had no respect in the community, that the reason was because we did not have the things that most farmers associate with success. Most farmers don't look at healthy animals, healthy fields, nutrient cycling and gross profit margin when selecting winners for awards and inductees into their groups. They look for the right colored equipment and plenty of it, a snazzy truck and a parking space for the banker.

In my experience, truly successful business people do not aspire to wealth. They do not say things like "I'm ready to be rich" when a sales opportunity comes along. For every promise that your unexpected ship is coming in, two ships you expected to come in don't. If I recounted all the times we've extended credit or tried to help someone get started, only to be stuck with the bill, you wouldn't believe it. But for some reason, people assume that successful business people got there by clawing their way up.

I'm sure that happens. All of us know business people who seem to be clawing their way. They round everything up. They figure and figure to make sure they get every last penny. They take every imperfect scale weight in their favor instead of the customer's favor. But they usually do not last long, and they do not build enduring business. What goes around comes around, and these folks get their comeuppance eventually.

Part of human nature is to assume that the only way people acquire money is by worshiping it. Or at least that it is ill gotten. This is unfortunate, because I think it can make us sabotage good business practices. We become so concerned about what some people will think if we actually make good money, that we hurt ourselves or our family. After all, "the labourer is worthy of his hire" (Luke 10:7). As we come full circle from establishing a healthy antipathy toward money, we must appreciate that it is the medium of exchange and the way we show respect.

Most farmers work for pauper's wages, and view themselves as society's bottom feeders as a result. Our self-respect is tied directly to how we view the value of our ideas or our time. No one can dictate those for us, but there is a balance between "getting all I can get" and "getting a fair return." You wouldn't believe how many people think selling a book for the price I do is highway robbery.

The only people who think that are folks who haven't bought one. Is that prideful? No, it's just following Dad's advice: "You may as well do nothing for nothing as something for nothing." He was constantly amazed at his farmer accounting clients who worked like dogs, but made no money. It's almost like an unwritten rule. You farm. You work like a dog. You earn peanuts. Why?

When farmers begin to view themselves with dignity, they will begin requiring better prices. It is not prideful to value yourself and your contribution to society. Even religions value themselves and demand service, respect and action. A big difference exists between putting my worth up to at least a mid-level attorney and putting it at a dishwasher. I mean no disrespect to dishwashers, but that level of respect is probably not the one that will put the best land stewards on America's farms.

Farmers can't possibly be the best we can be by walking around all day, head hung low, a "Kick Me" sign on our rear, humbly fulfilling our debt to society: grow food, feed the world, be stupid, be poor. "I don't deserve anything" is the mantra of false humility. If we do a good job, work hard, persevere and give a little better than asked, we deserve respect. The surest way to show that we appreciate the job we do is by demanding remuneration in kind. I didn't say demand to get rich.

If you feel like this chapter is confusing, it is for a reason. I'm confused. And that's the whole point. Money is a difficult subject at best, and one we never come down on right in the middle. It's not a proper goal, but it is a proper tool. Money is not something we strive for, but it is something we need enough of to pay the bills. If we're not making enough to pay the bills, we don't have a family friendly farm. And yet if we decide to be millionaires, that carries pitfalls galore.

In a nutshell, here are some closing thoughts about money:

- Money has nothing to do with contentment or happiness.
- The best things in life are free, but usually in business the most effective way to show appreciation is monetarily.
- My judgment of your materialism is highly subjective, and vice versa.
- I'd rather have less money than more if it will preserve my integrity, charity and proper perspective within my family.
- Not all people who drive BMWs are ostentatious, but it seems that way, so I won't drive one. Does that make sense?
- Driving a clunker is great and a pox on the person who faults you for it.
- If I owned a red Ferrari, two silver Corvettes and a baby blue Porsche, I would be materialistic. The way people perceive us does matter.
- Good farmers producing good food deserve premium prices.

- Businesses devoted to amassing big bank accounts will not be a credit to their community nor their family.
- I hate to lose money.
- I enjoy having enough money to give to worthy causes.
- I never want to appear mercenary, even though at times I think I am. But I hate feeling like I'm just doing something for money. It cheapens the whole deal. But I do that sometimes, and that's the truth. Unfortunately.
- I hate money. I love money. That settles it. Actually, maybe it does.

Chapter 4

Goal Setting

What are appropriate goals?

Every successful business should have written goals. Our goals keep us on track, heading us toward our mission statement. You should certainly have a mission statement. But somehow we often get bogged down in the goals. It is necessary to address this issue in a large context because family friendly goals are crucial.

Goals establish a climate for noble ambition. Too often materialistic goals cheapen our life's work, thereby rendering it too low in value to attract the next generation. Children enjoy striving toward noble goals. That is one of the reasons why we enjoy reading about great heroes, both fiction and nonfiction.

Goals should contain our core values, not ancillary Wall Street objectives. Let me illustrate what may seem nebulous with some concrete examples. I will call these cheap or low-value goals:

- Earn a net income of $50,000

- Double our income at farmers' market
- Get 50 percent market share
- Raise 12,000 broilers
- Save $20,000 this year
- Double our customers this year
- Produce 15 bushels of premium grade apples per tree

I know to the casual observer, these all may sound like perfectly legitimate goals. But I would like you to compare them with another set of goals:

- Add one complementary enterprise this year
- Reduce customer negative comments to two for the whole year
- Knock one person-minute per chicken off processing time
- Drop from two-day to one-day paddock rotations
- Reduce chick mortality to less than 5 percent
- Increase Omega 3 fatty acids in eggs by 100 percent
- Turn dropped fruit into something of value

I don't know whether the differences are at first apparent, but if you notice, the first list is more materialistic and the second one has to do with quality, integrity, diversity. More values-based objectives. Children love to devote themselves to big causes, and money or production volume is never enough of a cause to warrant a lifetime ministry. That is why children are not impressed with millions of eggs or millions of chickens. At least not impressed enough to want to continue the farming methods of their parents.

Certainly exceptions exist to every rule, but in the main goals that touch the spirit offer grand opportunities. I've listened to countless business lectures at farming conferences about making extremely specific business plans. At the risk of sounding like I'm opposed to planning or sound business plans, I contend that many of the exercises are useless.

In a truly family friendly farm, we should dedicate ourselves to excellence and things will begin to take shape. If we establish a goal of earning $50,000 it tempts us to make shortcuts or run over people in our quest for the goal. The monetary goal is not noble enough. And this, I think, is what derails many high quality small companies. They begin with an excellent product or service, and then get stars in their eyes and begin cutting corners or running roughshod over neighbors.

And certainly this is what has happened in the organic empire movement. The whole certification movement is nothing more than a smokescreen to lock up market share for the initial big players. Creating another hurdle to "get in" creates the fraternity atmosphere and tends toward feelings of scarcity rather than plenty. "I've got mine" replaces the sharing and camaraderie that characterized the pre-certification organic farmer gatherings. Now it's market secrets and carefully guarded tours, trademarks and protective terminology.

Many people have asked me why we didn't patent the term "pastured poultry" or "salad bar beef," "Eggmobile" and "Feathernet." Look, duplication is the greatest form of flattery. I'd like "salad bar beef" to become a household term. It doesn't matter who invented it. The issue is to eat responsibly and differentiate truth from error. I enjoy getting farm brochures in the mail in which folks use these terms to describe their products. It's a hoot.

Holding things close to the chest is the way conventional folks have operated for a long time. It's the stuff of Wall Street that infuriates the populists and lovers of equal access. As soon as we create goals like money and market share, a whole system of thought kicks in that makes us compromise our core values. We begin viewing other people as competitors we need to badmouth or at least fear.

I can't believe how some people respond when a new guy joins the farmers' market with similar stuff. They're scared to death the new vendor will take away some of their business. Let's be bigger than fear. We say competition is good, and yet in actual practice, people want to jump into protectionism at the first opportunity.

One of my earliest memories on this score was probably 30 years ago when my Dad was trying to get the local Farm Bureau Federation policy development committee to adopt an on-farm sales exemption clause so that farmers did not need million dollar government inspected facilities in order to sell directly to folks who came on the farm. Of course, most of the members of this local farmers' group were Grade A dairy farmers, and several of them had been involved with the American Agriculture Movement tractorcades, voted libertarian and were quite radical. They opposed government involvement in education, housing, social security and practically everything but the military and justice system.

But when they realized that Dad's idea would mean that a homesteader with a family cow could actually sell that milk to a neighbor, they were incensed. Oh, that just wouldn't do. We'd have lower prices because of the competition, and then current producers would go out of business. "Why, that would open up the market to anyone, driving down prices. Oh, we could never have that. Some of these town job, part-time farmers wouldn't even care if they made a profit. They'd dump milk and put us out of business."

I was in high school at the time, and I remember being totally disgusted at these men. We had always milked about three family cows, and I enjoyed milking. I had already penciled out that I could milk 10 cows, sell the milk at regular retail prices, and make a healthy living on the farm. Protecting their own turf was far more important than granting me freedom to get a little turf.

I've watched this same mentality within the sustainable agriculture community. Approaching life from a sense of scarcity is truly a disease. After I appeared on ABC News Tonight with Peter Jennings I received numerous requests from people around the country asking us to ship them sausage and other meats. My stock answer was: "I'm not in the empire building business. You need to find someone in your own bioregion that produces like we do and patronize that farm. We don't need to be immersing everything in jet fuel. Buy from your neighbors."

Then they'd say: "But I don't know anybody here who does it that way."

"How much are you willing to invest to find someone? As much as you will invest planning your next Walt Disney excursion?" That always caught them blind sided. The point is that here at Polyface we've never been interested in market share or building empires.

We still practice a completely open door policy for farm visits. Because of the numbers, we no longer give guided tours, but if you want to come and see, you're welcome. We've been blessed beyond measure, and the least we can do is openly share it with other seekers. Besides, you never know when a visitor will patronize one of our for-profit seminars.

Recently I did a corporate in-house presentation on "Thinking Sustainably" at Nike apparel in Beaverton, Oregon. As part of this series and in keeping with their policy of giving back to the community, they asked me if I would do a free lecture at a college in Portland. I was the sixth over a two-year period of these millennium speakers. Of course, I was happy to do a presentation gratis since I was already there. My Nike hosts expressed their amazement that I would so quickly agree to do the extra presentation, noting that most of their speakers–even those in the sustainable thinking movement--demanded additional fees. How disheartening.

We can't throw money out the window, but neither do we need to worship it. Between those two extremes is plenty of latitude. My experience has been that if we devote ourselves to excellence rather than money, the money usually takes care of itself. And in our headlong rush to acquire materialistic goals, we lose sight of excellence.

That is why the second list of goals is more appropriate for a family friendly farm. They deal with excellence. Lower mortality, increased nutrition, greater health, more diversity. These things result in a more functional farm and, in turn, one that yields greater income for the same amount of effort. But these are not the goals I typically see when a business planner begins his PowerPoint presentation. What the business guru shows is building a marketing plan, surveying the market--I wish I had a nickel for all the marketing surveys I've filled out for graduate students trying to ascertain the market for pastured poultry or some related alternative item.

Here's the point: if the effort expended in doing all the marketing spreadsheets and analysis were devoted instead to just doing an excellent job, the farmer's passion, coupled with the product's quality, would open doors of opportunity never realized in the market survey. One of our goals is always to allow the plants and animals to fully express physiological distinctiveness. As a result, our refinements yield a product that continues to get better and better.

Normally, the quality consistently slides once the product is successful. I know a fellow who grows Thanksgiving turkeys for one of the large organic foods supermarkets. Years ago, when he started, he ranged the turkeys on pasture. As the flavor and distinctiveness created market demand, the orders increased. He was running around trying to get the birds bagged and processed, compromising on the production end. Now he grows them in a regular confinement poultry house just like the industry, except with more square footage. He says allowing the turkeys to be on pasture was just too much work. But look what happened to the product quality.

Of course, when this happens, it just creates more market share for me because now mine are far superior to his. But my market share has far less to do with my marketing than it does with a continual pursuit to place the turkey in its most natural habitat. That kind of goal yields a nobility of mission that keeps a farm on track in the values that really matter.

A goal like creating another salaried enterprise for a child is far better than a goal of increased farmers' market sales. If your product is consistently better and your countenance more radiant, you will consistently see sales increasing. But with a goal of increased market sales, you will be tempted to talk down the competition, or to view the competition as the enemy rather than a friend. I know these differences may seem subtle, but I've watched the effect that the wording has on people, and it is profound.

What are your goals? That is the fundamental question. If you are grasping for materialistic confirmation as a measure of success, you will come up empty handed in the things that have the highest value. If, on the other hand, you are grasping for high quality, relationships and customer satisfaction, you will enjoy the most satisfying confirmation of your efforts. The money will take care of itself.

Through many experiences, I have learned to never put much trust in a man who swaggers in and says: "You just grow it and I'll sell it. You can't grow enough to satisfy what I can sell." Mark that guy. He has a dangerous attitude.

Beware the kind of goals you make. Put real emphasis on wording them so they really express your most heartfelt values. And word them so that they are noble enough to not only justify your life's work, but that of your grandchildren as well. Financial goals are cheap and unrewarding in the long term. Values goals, on the other hand, are worthwhile and tend toward multi-generational striving. That is noble.

Chapter 5

Appropriate Size: Growth has a Downside

The business success we've enjoyed with our family farm has made us wrestle with the question of appropriate size. Can a business grow out of its family friendliness?

This is a critical question because in a capitalistic society no growth equivocates to stagnation, and that generally spells doom. The family business books I' ve read spend a great deal of time dealing with all the legal contortions to maintain growth as well as family ownership. I realize that with the "farm" element in this book, we probably don't need to deal with businesses the size of Ford Motor Co., but I think it's healthy to at least bring up the discussion. If small businesses becoming large businesses are axiomatic, we'd better deal with the big business family friendliness.

And if big business makes it family unfriendly, we'd better recognize that not only before we get there, but before we even aspire to it. In our modern economy, the unspoken law is that a successful business grows its gross income at least, and net income preferably, every single season. If one or both of these doesn't grow, the year is considered an unsuccessful one.

One of the most fascinating elements of the whole back-to-grass movement in animal agriculture is the different physiology of the most pasture-adapted animals. Acutely apparent in dairy animals, these differences show up in rumen size, leg structure, even width of the mouth. Animals feeding pre-cut feedstuffs at a trough can get along fine with a narrow face and mouth. But grazing dairy cows need wide mouths to efficiently tear off and masticate forages. While the confinement operations shoot for production, production, production, and often have cows milking more than 20,000 lbs. per year, the grass-based operations select for efficiency and find that profitability peaks at about 10,000-12,000 lbs. per year.

The point is that the animals in the two systems look completely different. This close environment adaptation is normal, and explains why the common sheep breeds are all named after counties in England. Essentially, each county developed a breed most adapted to its area. Form follows function. Note the difference between heat tolerant Brahman cows in the south and cold tolerant, long-haired breeds like Galloways and Scottish Highlanders in the north. They work because they are adapted to those areas.

My question is if the same kind of thinking holds true for family friendly business models. Does a family friendly business have a size limitation, or is it size neutral? Obviously we've articulated many aspects of its physiology in the pages of this book. We've talked about techniques and models that make up this family friendly business model, as it were. But is this model size neutral? This is an important question because it speaks directly to aspirations, goals and vision.

Obviously, a business employing six grandchildren is larger than a business employing only the founding couple. Let's begin by looking at the philosophy of growth.

Growth is neither inherently good or bad. If a baby doesn't grow, we know something is wrong. But never forget that cancer is growth, and none of us would want cancer to grow. We all look at economic growth, expecting an increase in Gross Domestic Product. I don't know if that's good or bad, but our culture sure expects it. But we do not look for the same growth in murder statistics or highway fatalities.

These are obvious agreements across society. Some growths are not as clearly defined. For example, when our county board of supervisors increases the local school budget, the headlines blare: "Supervisors Opt for Growth." The editorial assumption, of course, is that the higher budget is a good thing. Those of us dedicated to home schooling specifically and limited government generally would not agree that this growth is helpful. We could take half the $8,000 government cost per child, distribute it to parents to spend at their neighbor elementary teacher's six-child home-based school. The teacher could keep her own toddlers at home instead of daycare, could eliminate the commute to work, get rid of the second car. The neighbor children would be close to home, in a much safer environment, and the teacher-student ratio would be incredible. Instead of duplicating eating and classroom facilities, students would utilize existing homes with their kitchens and internal rooms. Everyone would be better off.

But this kind of thinking is completely unacceptable to the powers that be. The only solution to education as they see it is to throw more money at it. They are operating with only one tool in their box. As the old saying goes: "If the only tool in your box is a hammer, every problem looks like a nail." And so their Procrustean solutions continue year after year. In Greek mythology, Procrustes was a giant who raided surrounding villages and brought prisoners to his hillside of iron beds, where he chained them down. Because all his beds were the same size, he cut off the legs of tall people and stretched the short people to make them all fit his beds.

Albert Einstein said "we cannot solve a problem with the same thinking that created it." But our government-educated civil servants have been taught only thinking that perpetuates the system, insuring that only solutions will be proposed that fit the prevailing paradigm. That is why government can never lead innovation. A political entity must satisfy 51 percent of its constituency, and 51 percent of the constituency by definition cannot be innovative because innovators are inherently different than the majority.

Growth is a relative term. That's the point. The question is what should grow and what should not. And it is unquestioned within American business that growth in gross income is positive. But growth can occur in areas hard to quantify. I well remember a sage farmer telling me he could make more money on 20 healthy cows than 30 unhealthy ones. I'd rather have one healthy 60 lb. calf than a truckload of dead 100 lb. calves. I know it militates against everything we've been taught, but somehow, deep within our souls, we must get rid of this love affair with statistical growth as an inherently wonderful thing.

This year Teresa and I took the lowest income we've taken in years. Most people would say that's terrible. After all, we're in those critical middle-aged years when we must be running up the nest egg, right? The reason for the low amount is that the children are getting most of it. What would Teresa and I do with a bunch of money anyway? We don't want to go on high falootin' trips, we have no desire to buy designer clothes or join the country club. Watching the children take over the farm and enjoy its fruits, invest in their own enterprises and begin talking about long-range projects is certainly growth, but it doesn't show up on the balance sheet.

Okay, I think we've settled that growth is not necessarily good nor necessarily bad. Other values pass judgment on the growth. Keep that in mind whenever you read about the next agricultural breakthrough that allows us to grow a faster chicken, a bigger pig or taller corn. For every action, there is an equal and opposite reaction.

I think the most important point I can make in this discussion is that you will never find easy street. Not that you would want to. But you will never "arrive." I have always said that the day I feel like I've arrived, just go ahead and put me on the compost pile because I won't be good for anything. Anyone who thinks our family has "arrived" has not a clue about the new challenges and pressures of running a larger operation. I remember a couple of years ago when I realized our farm gross income put us into the elite "large farm" category according to the USDA. I also remember when we graduated from "small" designation to "middle sized" designation. Having been in all three, I can tell you each has its own problems. Early on, our goal was just to survive, to pay the bills. Then it was to make some long-term capital improvements in infrastructure and landscape, like sheds, ponds, water system and front end loaders.

The final step came when we added apprentices. But then came the use of subcontractors. Then an agent for farmers' markets. It never really ends. I've said over and over that it takes just as much creativity and energy to operate this phase as it did to develop the prototypes in the first place. Now instead of just moving broiler pens and cows around a field, we have cows, turkeys, eggmobiles, broiler pens and sheep. On a family friendly farm, you never get done refining, trying to do better.

And more people involved with the operation create more opportunities for misunderstandings and relationship problems. In addition to the four white collar salaries generated directly on our farm, it also supports others in delivery, brokering and subcontracting. Then add all the informational ministry, and you have a fairly sizable business. Holding everything together is just as challenging as getting enough money together to pay the electric bill was twenty years ago.

To be sure, my job description has changed dramatically, along with my waistline. But it is no less stressful and not any easier managing, administrating and handling the desk work as it was to hand shovel 10 tons of compost a day when I was 25 years old. The trick is to maintain the same cohesiveness, same sense of teamwork, same appreciation for everyone's niche, as Teresa and I had when we began under Mom and Dad. Then it was just four of us. Today it's a dozen.

The bottom line is that growing the business has a downside. You need to decide what your family can handle. You will find that as it grows, your comfort level will grow too. Listen to your spouse, listen to your kids. Growth can occur in wonderful non-quantifiable ways and we must let that happen.

As more and more people become dependent on the business, the stakes get higher. The chances for someone to develop inappropriate goals, to "take the money and run," become greater.

Smallness has its beauty. Simple elegance. We have constantly turned down markets that offered big bucks, simply because we didn't want the stress of keeping up that part of the enterprise.

We still do not cut up chickens. We know we could sell more that way, but we simply don't want to do it. Period. Maybe someday a grandchild will come along who loves to do that. Fine, that can be his project. But the growth will be limited to the passion areas of the individual. And that's far different than having an appetite for growth, jumping into each new growth opportunity, just so we can increase annual gross income.

Staying small offers flexibility. A speedboat turns easier than an aircraft carrier. Resist the urge to be big--don't dismiss bigness out of hand, but resist the desire to be big. Let the growth come in small increments as you're able to handle it. Organic growth looks different than grasping growth. If growth comes in spite of your desires, fine. Desire the right things, things of lasting value, and your size will be just right for you.

Appropriate Size: Growth has a Downside

Chapter 6

Making the Break from Outside Employment

One of the most frequent questions regarding the whole family farm goal is about making the break from outside or supplemental employment. This generally comes from families who are already farming on owned or rented land and have a feeling of proficiency but are in those awkward in-between years.

Teresa and I faced this question back in 1982 and, while I don't purport to have all the answers, I think I have some thoughts that would be helpful. Obviously, when we sever that steady paycheck it really lays our financial situation on the line. That steady paycheck really is a safety net and not having it can shake up even a strong marriage.

The old song about self-employed people work 80 hours a week for themselves so they don't have to work 40 hours a week for someone else is laughably accurate. Many families have that yearning to work for themselves, and I know how they feel because we're one of them. Welcome. But it's not easy.

When Teresa and I were married, we lived cheaply. Real cheaply, and saved. With our own garden, meat, milk and eggs and living rent-free in the attic of the farmhouse, we saved almost 50 percent of my newspaper reporter paycheck. Anyone who has worked as a reporter knows the salary will not put you on the short road to wealth. But we had a dream for which we were willing to sacrifice. After 18 months we had $5,000, which we knew was plenty to live on for a year.

But what then? That's always the nagging question. What happens when the nest egg runs out. Then what?

This was the question facing me. We made the decision early on that Teresa would run the household. Two incomes were not acceptable. Nor was my staying home and her working in town. I remember well when the lights went on. As a reporter, I became friendly with many local businessmen. I was in and out of construction projects, government offices, the police department. I was "in the know," as they say. And what I saw was an aversion on the part of most people to do any manual labor.

I enjoyed working hard and thought physical labor was honorable work. As I did stories about local businesses and the regular happenings in a small city I became acutely aware of just how desirable I would be as an employee--to anybody. I knew how I hustled at home--and at the newspaper--and it was obvious the average person knew nothing about hustling. The average person was just "putting in time."

Suddenly it dawned on me that I could work anywhere anytime. In fact, I would be a plum for anybody. Here's why:

- I loved to work, and work hard.
- I was honest. No pinching from the boss.
- I was strong. Carrying water buckets had made me strong as an ox.
- I was likeable. I could carry on a conversation with anybody and was interested in just about everything. I could make people feel good about themselves.
- I followed orders. I wasn't a back-stabber. If I had a problem with a boss, I went to the boss and talked straight out. I wasn't big on backroom gossip parties that infect virtually every place of employment.
- I was creative. I invented a couple of things in the newsroom that made me twice as productive as the other reporters. And of course, instead of adopting my procedure, the others made fun of me. Oh well, so much for the guy who sticks his head above the rest.
- I was dependable. I'd break my neck to be on time or fulfill my commitments. Old faithful. That was what I tried to be.
- I was aggressive. Advancement was my middle name, and I bucked for every promotion and every difficult assignment possible. No lethargy here.

This is not an exercise in self-promotion or pride; this is just what I consider to be normal, good character qualities. In short, I realized that I would be an asset to anyone, and I could fit into a multitude of capacities. I could learn quickly, even if I didn't know anything about the job today. In fact, I was incredibly hireable.

When that thought hit me, it was like a ton of bricks falling off my back. Suddenly, I was liberated from my paycheck. I began watching help-wanted ads in the classifieds. The thought struck me that I could hire on anywhere as a dishwasher, and would probably be assistant manager within six months. That's not prideful, that's just how deplorable the quality of American workers is in this day and age.

Talk to anyone in the personnel department of a major business and you will hear unbelievable stories. I recently talked to the manager of a huge beach hotel dining room who said she couldn't get dependable help. She noticed a couple of her busboys missing during a banquet and went looking for them. Even though they had been there at the beginning of the banquet, they were nowhere to be found. Nearly two hours later, they showed up, dripping wet, and began bussing tables. Furious, she asked them where they'd been. "Oh, we just heard the surf was perfect so we had to check it out. The boarding was awesome," they replied. No apologies. These were college guys, trying to earn their way through school.

The dependability of employees, especially in the lower paying jobs, is nonexistent. And good ones don't stay on the bottom very long. Mediocrity is everywhere. And so is aversion to manual labor.

I have no sympathy for college graduates who say they can't find employment. I had a guy call me one time wanting to be an apprentice. He said he'd graduated from college eight months before. I asked him what he was currently doing, and he said: "Oh, I've just been sitting on my rear end." Guess what. He didn't get in. People in the trades are dying for good, dependable labor.

I realized it was not beneath my dignity to push concrete, shovel dirt or wash dishes. And if I had to take a job like that, I'd be the best concrete pusher, the best dirt shoveler, and the best dishwasher. As a result, I was free. Totally free. If we ran through our savings, no problem. I'd hire on doing anything anywhere long enough to save up some more.

But if you aren't willing to wash dishes, if that's beneath your dignity, if you need a white collar salary or nothing, forget it. Then your safety net just fell away, because you don't have the humility and character to get free of your golden chains.

Self employment is not for everybody, to be sure. It has numerous pitfalls. Among them are these:

- You take full responsibility. No "stupid bosses" to blame.
- The stress of the business is with you all the time. You can't leave the job at the office or at the work site. It's with you 24 hours a day.
- The family can grow tired of everything revolving around the business. "Don't we have a life?" can be a question the children ask, and it will be a red flag that you're not allocating enough family time.
- The average family business only lasts 24 years, the productive life of the founder. Not a good track record. Failure is the way to bet. Farming is no different. Look at the deed record on the average farm, and you'll see multiple owners, many of whom only stayed a few years. Our own farm had three owners in 12 years before we came in 1961.
- Taxes, insurance, retirement--now everything is up to you.
- Government paperwork. This is enough to keep many people away from self-employment today. Safety, labor, environmental and a host of other bureaucrats require reams of forms to be filled out, bleeding off your life energy.
- Unsteady paycheck. This is a biggy. Weaning from the outside job is tough emotionally and economically.

I don't know anyone who moves to self-employment for money. Most do it for personal satisfaction and to do something bigger with their life. That is why I stress having proper goals and thinking like a minister. Otherwise, the passion will not carry us through the reality checks that will inevitably follow our decision to break with outside employment.

Having a humble spirit and good work ethic cushions the break. It provides an emotional safety net that can become an economic safety net at any time. Having been there, I encourage folks who really want to go full-time to go ahead and do it. If it fails, at least you will have tried. But if you never try, you will never know if you could do it or not, and that will be a nagging question.

In our own case, our farm losses dropped precipitously when I made the break. I was here for every calf birth. In farming, timing is everything. The slippage stopped because the timing was better. Fewer crises. Increased efficiency, and projects just came up to help us make ends meet. The money was close, but it was always there. Slowly, month by month, year by year, we made progress. Not until about the fifth year did we really believe we were out of the woods.

It may take that long for you. But if your heart is in it, and especially if your spouse and children are in it, you can weather the tough times. And if you have to flip hamburgers for a month, so be it. But having made the break and being able to devote full time and energy to the farm, you will make progress so much faster than being torn and limping along part-time that your total time being part-time will be greatly reduced.

Be encouraged. Be committed. Be willing to do anything. You'll find making the break does not have the sting you may have thought. Go for it.

Chapter 7

Restoring Community

One step beyond the family-based paradigm is the community, an extension of our overall theme. Working in community is not something we Westerners are accustomed to doing.

Our culture has spent roughly a century going through a historic learning curve, developing incredible proficiency at a model with these basic tenets:

- work occurs away from the home, even away from the neighborhood
- people with narrow expertise dispense better advice than generalists
- hire it done, preferably with money--trust the experts
- my highest satisfaction occurs when I "do it my way"
- rugged individualism--"I don't need anybody, 'cause I'm a self-made man"
- independence from the community--homeowner's insurance, auto insurance, health insurance, life insurance, pension, Social Security, Medicare, etc.
- happiness and self-worth come from what we have--peer

dependent self concept

These are the pillars of our current cultural paradigm. A paradigm is a mental road map, a soul-level filter through which we see everything that touches our lives. The insidious thing about a paradigm is that it is so much a part of us that we do not realize we have it. And when two people with opposing paradigms interact, conversation is practically impossible. If my educational paradigm assumes direct parental and family-oriented life experience, a conversation with a person whose educational paradigm is a peer-oriented, salaried-expert, non-family experience can be an exercise in futility.

To illustrate this point, when I speak at alternative agriculture conferences I ask for a show of hands to the following question: "How many of you have ever argued a conventional farmer into an alternative approach?" I have yet to see a hand. The reason is because our hearts filter what we allow our heads to believe. I can present all the evidence in the world, but if you don't want to "see" it, you won't.

The point of all this is that the above cultural paradigm permeates most of us to such an extent that we cannot conceive of the learning curve wreck when we switch gears. The initial euphoria over a change can quickly turn to panic as the realities of a new paradigm begin slapping us in the face.

Almost no one entering community-based models as a natural extension of family-based models has spent much time practicing a community-friendly paradigm. In my opinion, here are some of the required tenets:

- work as close to home as possible; preferably in the home

- personal responsibility for health care, elderly care, education, etc.--I must become knowledgeable and make decisions
- help one another without keeping records of who got the better deal
- community responsibility and interdependence for insurance-type calamities
- true servanthood--I'll take the poor house site; you take the better one
- my happiness comes from seeing you prosper--I'm not an empire builder

Most Americans have been immersed in thinking and practice directly opposite this community-friendly list. To embark on this new path requires an individual learning curve. Add to that a whole bunch of folks going through it at the same time in close proximity and chances are the wreck will be compounded. We must understand that the Amish have a multi-generational legacy, a cultural paradigm, that bolsters community. For example, when faced with a decision, the Amish ask the question: "How will this affect our community?"

Honestly, I don't really ask that kind of question. I just assume if it makes my job easier, or if it will make me more money, it's okay. To ask the question: "How will this affect our community?" has profound implications on how I make decisions. Whether or not to buy a car, add a cow, dig a well or cut a tree--even how to cut my hair, perhaps--are directly impacted by my mind set, my paradigm. I'm not suggesting that the Amish question is the best; all I'm suggesting is that we twentieth century Americans approach problem-solving and decision-making from a fundamentally non-community paradigm that is so much a part of our being that it is absurd to think that we can quickly undo a century of baggage.

While it is true that a crisis can pull people together, after the initial crisis, life goes on, and what then? Now we actually have to put up with that weirdo next door. Every learning curve has its honeymoon, the euphoria period. Then comes the spat, the first big disagreement--the wreck. Working through that period is what gradually turns the thing upward and toward smooth running proficiency--real competency of a "simple solution" can reduce sound judgment.

My dad wanted to build community here. I picked up the same desire. For decades our church groups have been more interested in making sure baptism was performed technically correct rather than taking care of single moms or our own elderly. We had doctrine down pat but retirees frittered away their wisdom years tooling around in RVs playing golf and visiting the opera. We knew Scofield's meaning of each New Testament parable but sent kids away to public school to receive a good enough education to get a job a thousand miles away from home that paid well enough to hire nursing homes to take care of Mom and Dad. We were so busy intellectualizing spiritual gifts that we became physical pharmaceutical junkies.

We have been in this place since 1961, yearning for community. In just the last five years, it is beginning to develop. We schedule workdays to help each other. We actually have to require each family to submit a project because our cultural paradigm considers it un-American to ask for help. See what I'm saying? We've cut trees, demolished buildings, built barns and brainstormed landscape plans.

We've also created an index of expertise to stimulate barter and brethren patronage instead of exporting all our dollars outside the group. In the average Amish community, every imported dollar turns over seven times before leaving. Patronizing our own in that way is a mark of true community. If you think this is a great idea, let me give you a test. When was the last time you shopped at a Wal-Mart because the item you were looking for was cheaper there than at the small business? You see, here again it's easy to speak great swelling words for community, but most of us really do not operate that way.

In our group, we do not live on the same land. We are scattered around the area. But we have now done some projects for neighbors of our fellowship, creating a wider community respect and visibility. We've had folks come and go. Plenty of pain has accompanied gains. But our family is committed to community in all its social, accountability, economic, and outreach ramifications.

10 Commandments for Making the Kids Love the Farm

Chapter 8

Commandment 1
Integration into Every Aspect

Children need to be integrated into a family friendly farm in a meaningful way. They should be incorporated into the farm business as early and heavily as possible.

Children in diapers need to be involved with the activities. I have no idea how many times our kids fell asleep in my arms on the tractor seat. I'm sure someone will jump on me about safety here, so let's talk a little about safety to get it out of the way.

Life is risky. You can die from it. Can you imagine a life without risk? Who would climb mountains? Who would dream of flying an airplane? Who would sleep under the stars? Who would venture out the front door? The simple truth is that a life worth living is filled with risk. Anyone who wants to eliminate risk from his own life or the life of his children does so at the cost of taking the zest and fire out of life.

The biggest risk in the world is becoming emotionally

involved with someone--loving unconditionally. Who would get married if we were all afraid of emotional death--divorce? Risk takers get more out of life, period.

That doesn't mean we should be foolhardy. A fine line certainly exists between recklessness and boldness. I understand that. But we have become an incredibly paranoid culture, suffering a victim complex about everything. Which is more costly in the big picture: having one or two children injured in tractor accidents or losing all the farm children to cities?

Certainly a balance exists, and I don't advocate teaching swimming by the throw-them-in-and-walk-away method. But prima donna children are a disgrace to any parent. And overprotective, hovering, ooooohing and aaaaahing parents who never let a child fall in a cowpie or tumble down a haystack are handicapping a child for life.

The rough-and-tumble part of life teaches wisdom and caution. One of the things we've noticed since our children were tiny was their caution compared to the recklessness of non-farming children. Full of bravado and foolishness, these city kids have no respect for machinery, animals or falling trees. When we host field trips at the farm I'm always amazed at kids who jump off a moving wagon, or worse, jump between the tractor and the wagon before we've come to a complete stop.

Guardian angels must live in barns. Hardly a barn is immune from rope swings, hay forts and any number of acrobatic setups. And many a youngster has a broken arm or worse from the roughhousing that goes on there. But who would want to deny youngsters the chance to play aggressively? A foot of hay strewn on the floor can provide a cushion for falls. The wise child will see the danger in things, and the wise child has normally developed wisdom by being kicked, or having fallen at one time or another. Allowing kids to take some small tumbles early on gives them a precautionary sense when they get a little older. Overly protected kids do not develop that common sense.

Overprotection is as detrimental as underprotection. We dare not shield our children from reality to the point that they have no sense of consequences. The earlier we expose them to potential danger, sickness and tragedy, the easier they'll be able to cope.

We see this routinely in our customers who come to the farm to pick up their chickens. Occasionally customers call and ask if it is okay to come early with their children and see the processing (euphemism for slaughtering). The parents are generally the ones that turn a little green around the gills. The children are just fine. Eleven-year-old boys are the best. Daniel gives them the killing knife and lets them slit some throats.

These boys get a real thrill out of this. They see real blood. And the slice is final. The chicken doesn't revive like it would on a video game. It doesn't have a bank of lives on which to draw during its numerous mishaps. Furthermore, the chicken doesn't talk to anyone.

It doesn't offer a Disney story on its way to finality. It just looks around, finally gets lightheaded, closes its eyes, and fades away. When Daniel was in diapers he would drag the killed chickens across the yard to the scalding tank. His little fat bottom waddled along as he struggled with two birds nearly as big as he was.

But by the time he was 10 years old, he was a master

executioner. And now he has taught his excellent technique to countless others around the world. A lady and her husband were here for a couple of days learning the processing techniques. Daniel was about 14 at the time and he had her down close and personal with a chicken's neck, showing her the proper jugular slitting procedure.

Toward the end of the lesson, she laughed: "This is what I always wanted. To turn 40 and be taking orders from a 14-year-old."

What if we had decided this was too gruesome for a child? What if we had decided to shield him from the realities of life? Would he have suddenly jumped into it as a 10-year-old? No. He would have developed other interests. At the very least, he would have viewed the procedure as ishy gishy and been sissified about the whole affair.

A family friendly farming model, in order to incorporate the children early, needs to minimize heavy metal, unhealthy systems and potentially dangerous procedures. This may be one of the best reasons to farm without chemicals. We do not apply anything to the soil that our children can't eat . . . in reasonable quantities. We need not worry about them stumbling into a room with a skull and crossbones on the door and ingesting some devilish concoction.

Farming is statistically one of the most dangerous occupations. But the danger comes primarily from chemicals, manure lagoons, silos and heavy machinery. One of the primary prerequisites for a family friendly farm is that it be child friendly. The average industrial farm is so child unfriendly that the adults have to keep the children away from much of the work in the early years.

Then when the children turn 13 or 14 and could really be useful, they've already developed interests in video games, little league and ballet. It's too late to awaken or create a love of farming. Designing a child friendly production model requires moving 180 degrees away from the industrial model. What parent wants to send a five-year-old up to the concentration camp factory house to check on the animals? That house is as inhospitable to people as it is to the animals.

The ammonia fumes destroy mucous membranes in the respiratory tract. The fecal particulate in the air coats the skin and clothes. Disease problems abound. That is why 75 percent of every conventional farm conference deals with diseases. Nothing destroys the desire of an aspiring farmer, regardless of age, faster than constant sickness and disease. I wish I knew how many children decided to leave the farm after the second walk-through the poultry house picking up dead birds.

That confinement house is full of whirring feed augers, manure belts, ventilation fans and countless other sophisticated contraptions that a child can't begin to understand. But our pasture models have no motors or clothes-tearing augers. Even a child can safely go out and check on things or fix whatever is not quite right.

When Rachel could barely walk she would go out to the chickens with me and move the water hose from bucket to bucket. We carry water out to the field in big tank trailers. A plastic pipe can swing in an arc big enough to fill several empty buckets at a time. A tiny child can pick up that pipe and swing it from bucket to bucket. The water comes out just as fast for her as it does for me. Farm models that allow the children's time to be just as valuable as the adults' will yield unbelievable returns, not just emotionally but economically as well.

Models that create room for children with adult-level, meaningful jobs will create the self worth and team player mentality that will fire a passion in their soul for this kind of work. Relegating children to menial tasks or make-work is a sure way to blow out any flame they may have.

Children can spot "make work" a mile away. If you can't think of anything for the children to do, then you're not operating a family friendly model. Instead of using a manure lagoon, we do large-scale composting with pigaerators. The pigs eat fermented corn in the bedding and aerate the bedding pack as they churn through seeking the corn. How many children and adults have perished in manure lagoons? But this pig-powered composting procedure offers a wonderful child friendly experience.

Not only is watching the pigs turning the bedding great fun for a child, but it is something that can be enjoyed with Mom and Dad. Compare that to the pumping and spreading and scraping of a manure lagoon, and you can see the difference. The pigs are wonderful entertainment.

We use good-sized pigs to do this work, letting them grow to more than 300 pounds. Daniel learned to slip an empty 5-gallon bucket over the snout of a pig and hold onto the bale, riding High-Ho Silver style. Who needs to get thrills at a beer-sponsored theme park when you can rodeo on a 300 pound porker across the compost pile?

When Dad has to don a space suit to go out and spray the crops, the model isn't child friendly. The smaller the machine, the more child friendly. I'll never forget the first time Daniel drove the tractor. He was about 6 years old and we were collecting bales in a 13 acre flat field. He had a whole field to turn around in and he was going slow enough that if he got into any trouble, I could run up to the tractor and jump on. He knew where the ignition switch was and there literally was not a place where the tractor would roll of its own accord.

He was so proud of himself that he couldn't wait to tell everyone at church the next day what he'd done. In fact, it was the first and only time he ever complained about being homeschooled. He felt cheated out of the opportunity to brag to more people about his accomplishment. The tractor was simple and small, easy for him to operate.

Beyond the chemicals, equipment, and buildings, though, is another facet of this child friendly modeling that needs to be addressed. That's the people issue. In large scale industrial agriculture, many farm families do not feel safe sending their children out into the fields with the workers. The workers are not friends and neighbors, like the old threshing rings.

Instead, the workers speak a different language and come from a different culture. Believe me, I don't think I have a racially or culturally prejudiced bone in me. I don't even laugh at ethnic jokes. But when we create an on-farm environment from which we must shield the children, the farm fields become a foreboding place rather than a friendly, beckoning place. Children view the farm with fear instead of romanticism.

A corollary to this principle can be found in the condescending tone that many farmers take toward their employees. "The hired man" description often carries with it a tone of superiority. This may carry over from slavery days, but it is unhealthy for children. It bespeaks of work fitting the owner and work fitting only underlings. And that does not incorporate the children into every aspect of the farm.

If you teach a child that he's above mucking out the manure or pulling weeds and you'll raise an arrogant, uncompassionate hard-to-please master. Studying the life cycle of family businesses both experientially and in books on the topic has convinced me that this is probably the most critical area of multi-generational business. The reason the average family business does not outlast its founder is because the second generation grows up with the silver spoon and doesn't appreciate the detail, the nuts and bolts of the operation.

Successful multi-generational businesses almost always started the second generation as janitors, on the bottom of the ladder. This early willingness to start at the bottom and be promoted by performance only is probably the key element that separates short-term family businesses from long-term ones. But if the entire model is child unfriendly, especially in terms of the labor force, the child cannot begin at that point. And the early years of protection will create a spoiled attitude, guaranteed.

This is one of the reasons why huge conglomerate agriculture cannot be family friendly. A farm that encourages child integration will necessarily be scaled to a family size. Scale does play a crucial role in how much a child can be a part. Scale affects the type and size of the labor force, the noise, dust and disease of the entire operation.

Finally, children should be incorporated into the financial and business planning side, early. I marvel at the young adults in their 20s and 30s who have no idea what the farm expenses and income are. This information will not hurt children. I can't for the life of me figure out why farmers are so close-mouthed with their children about the financial matters, unless it's because most of them are embarrassed to reveal the truth about how poorly the farm is performing.

Teresa and I have been accused of listening too much to our children, as if making them think their ideas could affect us somehow undermined our authority and created disobedient children. From day one, our children have been perfectly at home talking to adults and befriending elderly folks. The reason is because we didn't talk foolishness to them as children.

We didn't give them nicknames that ended in "y" and we didn't "babyfy" our voices when talking to them. We used adult language, spoke to them as adults, and explained a word if they didn't know what it meant. The ideas the children brought to be table during planning sessions, even at the ages of 8 or 9, were nothing short of astounding. I think children have an amazing capacity to be creative and bring fresh solutions to the table if they have the assurance that their thoughts are valuable and will be given due consideration.

Talking impishly to children creates jaundiced, selfish young people who think they are the center of attention. I think it even encourages young people to act up, put on a show. After all, it is this doting, impish talk that makes them believe they are the center of the universe. Children will generally rise to our expectations of them. If we don't expect anything but rote obedience and submission, we won't get vibrant awareness and interaction on business matters.

They will conclude that their ideas don't matter and will begin to feel unimportant and unnecessary. Then they'll begin looking for a way out and Mom and Dad will wonder why the children don't act interested in the family business. And the children will become reluctant helpers instead of eager beavers. I would rather the children hear some things they maybe shouldn't than have them feel like outsiders and subhumans because of all the closed-door sessions and secretive planning going on at the top.

Children aren't stupid. They know when their opinions and input are valued. Fortunately, they are also resilient. That is why I think less damage is caused by them participating in some discussions that may be a little heavy than being denied ones in which their participation would be appropriate. If I err, I'd rather err on the side of letting them have a little too much information than not enough. They can sort it out later. But they will have a real hard time sorting out devaluation, condescension and disinvolvement. If they realize how tight things are financially, perhaps they will appreciate why they can't have every indulgence. Keeping them in the dark only causes confusion, mistrust, bitterness and frustration.

Whether you are contemplating a farm or looking at your existing operation, look at it through the eyes of a child. Identify the points where involvement is being discouraged and see if that can be restructured to a more child friendly technique.

For example, when Rachel was an infant and we were processing chickens, we brought the playpen out in the yard for her to play in and sleep in. It would not have been practical for her to crawl around the blood and guts on the processing floor. But there was certainly no reason to protect her eyes and ears from the process. And for Teresa to stay away would have seriously jeopardized the quality control table. Many families who have infants or extremely small children are hamstrung from doing anything productive on the farm. It won't hurt a small child to crawl around in the dirt. If you haven't used chemical fertilizers and pesticides, ingesting some soil will only build up the immune system.

That is much less dangerous than throwing a 4-year-old in a room with 30 others at a daycare center or school. Our culture is paranoid of children being exposed to death, sickness, weeds and work, but doesn't bat an eye at exposing these tender immune systems to the rampant viruses in a school child factory. Get real.

This whole integration issue is another reason I like small animals. A chicken can only do minimal damage to a child, as opposed to a cow. A diversified production portfolio will offer more variety and more chances to integrate the children. Instead of just growing an apple orchard, for example, add some strawberries underneath. Now while the adults tend the trees, children can weed strawberries. Some work stations are not child friendly. In those cases, create a symbiotic one nearby so that the children can still be nearby doing their age and size appropriate jobs.

The bottom line is this: do not compartmentalize your farm life. Realize that a family friendly model is an integrated whole, full of participatory entry points for children. Treat the children as partners, as fellow team players, and they will rise to the occasion. They will develop a sense of ownership and a powerful sense of self worth that will carry their involvement into the future.

Commandment 1: Integration into Every Aspect

Chapter 9

Commandment 2
Love to Work

Require and allow the children to work. And work with them. This goes above and beyond integration because it requires sweat. If involvement is only mental, children will be shortchanged the character building goodies that accompany physically sweating over a task.

Unfortunately, to many parents what I am about to say will be viewed as child abuse. All I can say is time will tell which view is right. My challenge is that if you think what I'm about to say is abusive, let's have our grandchildren get together and we'll see which ones are more productive, balanced, delightful folks to be around.

Spoiled brats are the norm in our culture, where children are worshiped. Practically any law can be passed if it can be presented as positive "for the children." We can justify building more kid factories (conventional schools), requiring mandatory vaccination programs and supporting more unwed baby factory moms by invoking the "for the children" mantra. We coddle and dote on these little guys and gals as if someone died and made them kings and queens.

Believe it or not, little people really want to grow up and be like big people. Early exposure to the routine of big people--most of which involves work--will yield eager beaver adolescents who love shouldering their part of the burden. Just as all work and no play makes John a dull boy, so all play and no work makes John a spoiled brat.

Work is where children learn perseverance and discipline. I remember my brother and I being required to chop thistles for an hour a day during the summer. I made a game out of it, narrating the running battle of the Salatin's against the Thistle Nation, as we moved around the left flank to encircle the enemy. And over the years, as the number of thistles dropped, we actually won the war. Our children haven't had to fight that battle, but refinements and progress insure that new battles can be waged. Multiflora rose has replaced thistles as a criminal of choice.

Believe it or not, work can be fun. Everyone thrives on competition. Turning a chore into competition will turn it from drudgery into a game. For example, when setting fence posts, Dad can use the tamping bar and 4-year-old child can shovel in the dirt. The game is to see if the child can shovel in dirt faster than Dad can tamp. Sound silly? No way. This is as basic as it gets.

When you go out and gather eggs, once you're assured that a child can be trusted not to drop them, then teach efficiency. If you have five nest boxes to gather, send the child to gather out of one and see if you can gather the other four. Whoever gets done first, without breaking an egg, is the winner. If either one breaks an egg, he's disqualified; automatic forfeiture. This teaches quality as well as efficiency.

If you're weeding a row of vegetables in the garden, don't let the child lolligag around chasing butterflies all the time (some of the time is fine). But say: "Junior (or Juniorette) come over and see if you can beat Mommy."

Roll your eyes dramatically and drawl: "I'll bet you can't beat Mommy." Usually that's enough of a challenge to catch the attention of any tuned-in youngster.

"Ohhhh, I bet I can," says Juniorette, playing the game. Portion off two feet for the child and 10 feet for Mom, and then do the whole bit as if a race were starting. Children love races, relays, whatever.

"On your mark, get set, Go!" And the race is off. Don't let the child win routinely. Make it a tough competition. Children can tell if you're throwing the race and they'll reduce performance accordingly.

Children love to help their parents. If you have a 10-year-old who complains about work assignments, you've already made egregious mistakes. I'm not suggesting you can't begin today to try to rectify the damage, but it will be extremely difficult. This is why I'm adamant about starting early, early, early. Our expectations of these kids and the foundation principles we lay early will bear fruit during the pre-teen and teen years.

When Daniel was in diapers I'd take him with me to cut wood or build fence. He'd take his toy truck or tractor and play in the dirt nearby. As soon as I had something he could do, I'd call him over and we'd work together. At first, he would "help" me shovel dirt. But that's where it starts. Even as a toddler he would throw a small stick into the trailer when we loaded firewood. He was learning to work.

Tragically, the huge industrial crop farm offers few of these work opportunities for children. These row crop operations do not even build fences anymore. You can drive across the grain belt and see old broken down fences sagging into the side ditches of roads for miles and miles. The animals are gone and it's just grain, big equipment and storage silos. In my humble opinion, work is not child abuse. What is abusive is that now all the children from these families have so little to do that they are emotionally and psychology ruined by growing up thinking that soccer and video games are what life is all about.

Then they get out on their own and are expected to act responsibly, care for a spouse and children, pay bills, show up on time, and they are suddenly deficient in the most salient requirements to be a happy, well-adjusted person. I am not suggesting that we have sweatshops, of course. But I am suggesting that a childhood illustrated by the Laura Ingalls Wilder Little House on the Prairie books produces better human beings than coddled kids growing up on a diet of Twinkies and Saturday morning cartoons.

When Daniel was about eight years old he would get together and play (during his few minutes of allowed relaxation--ha!) with a neighbor boy. As little boys are inclined to do, they decided to build a fort. They planned it out and then accumulated some posts and boards. Then construction began.

Daniel took over as posthole digger and began digging the

holes for the corner posts. The boy's mother called us later that morning and exclaimed over Daniel's perseverance. "He won't even stop for a drink. He's working on that thing like his life depended on it."

Know what? His life did depend on it. In fact, we should all approach our work as if our life depends on it. And the drinking part. Guess where he got that? From going out as a toddler to help me build fence. One of my techniques for personal performance is to refuse to stop for a drink of water until a certain number of posts are in. That way I push myself to work steadily. That in itself becomes a competitive event with a reward at the end.

Daniel would come toddling over and say: "I'm thirsty." Instead of stopping and getting him some water, I'd say: "When we (not I) finish this post, then we can stop and get a drink, but not before." No whining--it wouldn't do any good. The only reason children whine is because they've learned Mom or Dad will give in if they are persistent. As soon as children learn that their whining yields no results, they'll quit. Whining takes a lot of effort.

One of the best books on this subject is the classic Dare to Discipline by James Dobson, founder of the ministry Focus on the Family. The principles he articulates are timeless because they are truth. When I'm in an airport watching parents struggling helplessly with a twisting, turning youngster my heart aches for that Mom or Dad. They make their life difficult by refusing to deal with child behavior consistently and aggressively. Tantrums and whining are actually simple to control.

The problem is that most parents today are so unsure of what they believe that they pass on that lack of resolve to their offspring. The children learn to manipulate with pouts and tantrums well before their first birthday. They poke and push, desperately looking for their boundaries. When they find that the boundaries are negotiable, these kids grow up defiant, disrespectful and rebellious. Daniel is now 20 and Rachel is 15. We've been blessed beyond measure with two youngsters who haven't turned snotty or rebellious.

This is not by accident. It comes from having a right model. And a functional child rearing model requires time, consistency and love. When people talk about our kids being clones, I quickly respond: "Look, I'm not a great parent. But by being together 24 hours a day from day one, this family friendly farming model builds into the system a forgiveness that doesn't exist when togetherness is compromised."

The point is that when I've spoken incorrectly, I'm there right away to see my child's response and make amends. The same God who demands children to respect and honor parents commands parents to not incite their children to anger. It works both ways. It is just as wrong for children to defy their parents as it is for parents to create hostility in their children. By the same token, when the children have done less than what is right, Teresa and I have been right there to deal with them. The consequences have not been delayed until an "appropriate" time. The best and quickest way to deal with anything is up front and right now. And when we're together all the time, that is anytime.

The hardest thing in the world is to deal with a problem by time delay. The circumstances get fuzzy, the "he said" and "she said" get all mixed up in our memory and any consequences lose the benefit of immediacy. This is why the entire issue of cottage businesses is such a hot one right now in our culture. As parents lose their children in the value transferral arena, parents are crying out for a better way. This spawned the home schooling movement and now is fueling the back-to-home business movement, of which farming is only one type.

Work attitude reveals your child's heart like nothing else. A willingness, an eagerness to work, reveals the true character of the person. Show me a person who loves to work, and I'll show you a person who will be a successful person.

As I write this today, Rachel, at 14, is cleaning a house for a lady in town. She goes two days a month. She and some friends have started a little loose-knit girl's group called "The Ladybug Club" and their Thanksgiving newsletter featured a section in which members wrote the typical "I'm thankful for . . ." The items ran the normal gamut of birds, trees, flowers, parents, friends--the normal things you'd expect from a group of homeschooled young teens.

But then as I read Rachel's entry, I broke down, big tears rolling down my cheeks. She wrote: "I'm thankful for work. Think of the chaos in our world if it weren't for work . . ." We have never preached to our kids about work; we have never had to scold them about being lazy. It is as natural as the night follows the day that if our kids have a healthy experience with work as toddlers, they'll develop a passion for it as young people.

Another of those teary experiences came a couple of years ago when I sat down at the desk after Rachel had been on the phone. Some of us doodle when we're on the phone, and it's usually nonsense--not even words. Just coloring in the O's on a letter, building squares or whatever. She writes words. And here's what she doodled:

"God loves me
Mom and Dad love me
I love me."

Isn't that precious? Talk about good self image and being rooted. And this from a girl who, when asked about what's she's thankful for, says "work." I'm belaboring this point a little because I think many adults are under the delusion that for an adolescent to say she loves to work indicates mental and emotional derangement. Such a person would be viewed as being not of this world. Well, she's not, to be perfectly honest. And for that we rejoice. I don't want what I see hanging around the shopping malls. I get the shivers when I see these body pierced, purple haired, tent panted young people. What are we coming to?

And the newspaper is full of articles about nothing for teenagers to do. Oh, the bleeding hearts who must come up with programs, midnight basketball, skateboard parks, youth centers and the like to get these teens off the streets. After all, cruising hamburger alley isn't doing right by the future leaders of our community. We must have activities for these teens to do so they don't have to roam the streets at night.

My kids read these stories and cringe. Daniel's pet response: "Some people have just a little too much time on their hands." Our kids are heading for bed at 9 p.m. after a full day's work, play and study. Do you want to know what true child abuse is? It's not coming in exhausted from a day's work well done. It's growing up undisciplined. It's reaching maturity thinking the night is for partying instead of sleeping for the next day's important work. Child abuse is growing up thinking my enjoyment must come from what someone else will give me or do for me. That's child abuse, and our culture is guilty.

It's easy to get all bent out of shape about Pakistani sweatshops and female circumcision in Africa. The easiest thing in the world is to point fingers at other cultures. I guarantee that if our children were working in family businesses they wouldn't be out on the street at 2 a.m. getting stoned or sexed. Let's work on the log in our own eye before we work on the splinter in someone else's.

I now call Daniel my slave driver. In fact, last fall a project opportunity fell into our lap that I wanted to turn down. But Daniel said: "You know, I don't think we've been pushing quite as hard as we should recently." And because of his desire, we bit off the project and got it done. Some folks calls us workaholics. The folks who call us that don't get much done. Life is serious business. And work is the fundamental aspect of life. God commanded that we would work by the sweat of our brow. He didn't say we would have a life of leisure and a safety net at taxpayers' expense.

Work can be made enjoyable not only by turning it into competition, but also by keeping it task oriented, not time oriented. For many parents, this piece of advice will justify the purchase price of this whole book. One of the surest ways to destroy initiative is to teach our kids to punch the time clock.

Just as punching the clock in industry creates workers by and large who nonchalantly say: "I'm just putting in my time," so time oriented jobs around the home and farm destroy initiative. Examples are myriad:

"Weed the green beans for one hour."
"Practice the piano for 30 minutes."
"Do school work for one hour."
"Clean your room for 30 minutes."
"Split wood for one hour."

The list could go on and on, but I think you get the picture. Time orientation teaches children to dawdle. It does not teach them to get with it. Every job should be task and completion oriented. Nothing seems as hopeless as putting in time with no completion in sight. Clearly defining the task and setting a completion point encourages efficiency and aggressive work.

Instead of practicing the piano for 30 minutes, the instructions should be to get song A to a certain level of competency. Pulling weeds should be defined by area or row. That way if it takes 5 hours because the child fooled around, so be it. But when this technique is used consistently, it fosters the notion: "Okay, I'm going to hurry up and get this done so I can do something else."

Never, never, never give time oriented tasks. It brings out the worst in all of us. Imagine the difference between an assignment to read a textbook for an hour rather than to master chapters 4 and 5 by tomorrow. Which one makes you get up and going?

This is why I oppose prisoners "doing time" and why prisons have such a low success rate. When conservative politicians try to get my vote by promising to "put them in jail and throw away the key" they do not appear "tough on crime" to me. Instead, they appear hopelessly ignorant about human nature, human dignity and proper punishment. As recidivism rates show, jails have nothing to do with punishment. The new operative word in schools is "time out" instead of the board of education applied to the seat of knowledge.

It used to be if you were in trouble at school you were in trouble at home. Today, some school official lays a hand on a snotty kid and the parents sue. No wonder too many young people have no respect for acceptable behavior.

Now we come to the second part of this work assignment--what comes after. Children have a great capacity to work hard, but only for brief periods. Even adults accomplish far more in small bursts than a tedious day without breaks. The steel factories learned early on that workers could actually carry more ingots per day if given a 10 minute break every hour than if they had no breaks and worked slower. That's just the diversified biological process, of which humans are a part.

On a family friendly farm, the way to get the spurt of energy for the task oriented assignment is to give adequate reward upon completion. For small children this can be as little a thing as having a story read to them, or eating a cracker. It can be a star on a work chart. But if after completion of one task the only reward is to be given another arduous task, all the benefits of task orientation go out the window. All bets are off, because the incentive is gone. Now we're back to wasting time because the reward is more work. Great reward.

If the reward for finishing weeding the corn is to go immediately and split wood, the child will probably just stay in the corn patch fooling around. But if a trip to the swimming hole is sandwiched between the two, with only an ending time having been established, then adequate incentive exists to get done quickly with the corn. The reward can be time oriented and that's fine.

Actually, the best reward is one with a time on the back end but no starting time. If the recreation has an open start time based on task completion, then it operates to maximum benefit. "When you get done you can play until 2 p.m." allows the child to kick in the free time based on his own efficiency. Children learn quickly to speed up the work in order to maximize the play time. If you say: "When you get done, you may play for two hours," it does not give the incentive to finish early like the open starting time does.

To this day we buy ice cream bars as a reward given at the end of dressing chickens. It's just a little something to look forward to while we're working hard on a less than favorite job. We all know that at the end of the task waits a luscious ice cream bar.

Perhaps the most common complaint among parents, especially farming parents, is that their children work reluctantly and when they do work, it's not aggressive. These techniques will help insure that these poor work attitudes never develop. Never give the kids time oriented tasks. Period.

One of the biggest temptations parents face is to not let clumsy little tykes help during the formative years. "I can do it faster myself," we often mutter, and ease the child out of the way. This is child abuse. This attitude denies children the yearning of their souls: to be Mommy and Daddy's big helper. As a type triple-A choleric gold, I had to fight this desperately when our children were small. Waiting for those little hands to tie a knot or pick up a board, pound a nail or get the screwdriver into a Phillip's head can be torturous when we're pushing on a project.

Watching a little child fold clothes or agonize over measurements while trying to get a batch of Christmas cookies made before company comes can try the patience of a saint. But think of it as an investment. Just as an investment looks like a big expense today, we do it because of compounding interest and the big payback years down the road. This is the way we need to look at children's work. It's a big investment today, but very shortly--and

I mean very shortly--they will develop skill at the task and displace the adults at doing it. That's when the adults cash in.

Those extra few minutes the job takes when the children are infants can mean the difference between having a super child alongside you in your waning years or the agony of reaching 60 and not having even one child who loves the farm. That is what Stephen Covey means when he says to begin building with the end in view. Every day we must visualize that end, keeping it at the forefront of our mental picture for the future.

Work offers parents and children a bond that does not occur any other way. Years ago we had a 14-year-old on the farm for the summer whose family lived on an 80-acre farm. Daniel was 10 at the time and had much more common sense than the older boy. That summer I had the privilege of teaching this young fellow how to drive a pickup and how to use a digging iron. I felt sorry for his father, who missed enjoying his son's achievement. Watching the sweat pour off his face as he wielded that digging iron was more rewarding for me, I think, than watching the boy play basketball was to the father. It was more special because we were together in the great outdoors, both sweating over a common task. A dad sitting in the bleachers can certainly be proud of his son, but it's just not the same.

Parents, don't cheat yourselves into thinking you'll have the same bond over recreation and entertainment that you would over work. The emotional depth and the import of work is far greater than a shared recreational experience. That's just play. And please don't think I'm suggesting that we should not play or participate in recreation. I do, however, believe it is greatly overrated in our culture, and has tended to cheapen the value of the human experience.

As the children get older, give them more and more unsupervised, independent tasks. Let them be the boss. That way they learn firsthand about things like getting a pickup stuck in the mud and making sure all the corral gates are closed before putting the cows in. Experience teaches wisdom and common sense. There is no short cut. And if we can't trust these youngsters to execute tasks on their own, they will never develop the self confidence to solve their problems and take on new projects.

Work. Vocation. This is what identifies people. And how it is done creates the framework of our value system, our integrity, character and relationship with others. Children love to work as long as parents nurture that passion with proper attitudes and techniques.

Chapter 10

Commandment 3
Give Freedom

Give the children freedom to do their own projects and exploration. Tightly controlling what children are allowed to discover or perform on their own initiative will make them timid and uncreative.

Certainly freedom must be offered within the confines of prudence and safety. That is not the issue here. What I see routinely is farmers allowing freedom to do only what Dad dictates. Anything else is out of order. As a rule, I think about anything a child feels comfortable performing, he should be allowed to do. If you have spent enough time with your children as toddlers, they will grow up with a healthy respect for machinery and danger. Daredevils generally have not been exposed to danger. A youngster raised right will know where his comfort level is and stay within it, whether it's running the tractor or handling cattle.

The old songs "but we've never done that before" and its first cousin "we've never done it like that" have stifled the ambition of many a youngster. The new idea, the new project, for some reason feels threatening to the older generation. I do not know why this is unless it is based in a lack of self confidence dating back yet another generation.

Why should anyone feel threatened with a new idea? New ideas are the stuff of improvement, the foundations for refinement. The day I quit seeking new ideas is the day you might as well shoot me because I won't be worth a hoot. And yet family farms are loaded with children and young adults whose ideas and projects get quashed.

Why is it that the first reaction to a new idea or project is: "It probably won't work." In his wonderfully perceptive and readable book Paradigms, Joel Arthur Barker lists the phrases people use when touched with a new paradigm. Here they are:

"'That's impossible.'
'We don't do things that way around here.'
'It's too radical a change for us.'
'We tried something like that before and it didn't work.'
'I wish it were that easy.'
'It's against policy to do it that way.'
'When you've been around a little longer, you'll understand.'
'Who gave you permission to change the rules?'
'Let's get real, okay?'
'How dare you suggest that what we are doing is wrong!'"

Pretty right on, isn't it? The power of parents to either stifle creativity and initiative, or catapult it, is incredible. That is because parental approval is one of the most basic yearnings in the human soul. To be approved by Mom and Dad is the heart's desire of every child. It's one thing to fail a fellow employee; it's another to fail parents.

As a result, the fear of rejection can stifle the most excited child. I run into middle-aged farmers all the time who want to do things differently. It may be something as simple as where to hang buckets. As we talk, they begin unloading all this pent up frustration at not being able to make any changes. Dad and Mom are in their 80s, but the 50-year-old son finally concludes with the killer phrase: "Dad won't let me."

"You're a strapping 50-year-old with a married daughter and two strapping sons. What do you mean 'he won't let me?'" These are gut-wrenching discussions and they leave me emotionally drained. There are no easy answers in these situations. This lack of freedom produces a feeling of suppressed energy, missed opportunities, destroyed passion.

The young generation must feel a genuine acceptance on the part of the older generation to offer input and new ideas into the operation. Otherwise, the young people never feel a part, never feel ownership, and never develop a love for the business.

If kids don't feel like their comments are welcome, they won't offer any. This happens between any two individuals, but it is most acute and most devastating between parents and children. Intergenerational discussions can be the most awkward exchanges imaginable. The piercing silence within the deep discussion is pregnant with unspoken hurt and pain. Mom and Dad, stalwart and rigid, try to make the best of the situation. The child--age doesn't matter in this scenario--breaks inside with unspoken anguish: "If I could just tell them how they make me feel."

But it's too painful, and the exchange breaks into the weather or soccer practice, something less threatening. Something that years of experience has defined as a mutually agreeable conversation within the confines of the family's freedom. This is not freedom. It's bondage. It's like saying that in jail you have freedom to get a meal and go to the bathroom every day. That's not freedom. And yet it is a metaphor for the level of freedom permitted in many family businesses.

For all my imperfections, I want my kids to be able to say: "Dad freed me up to pursue my projects." Each of our children is different, and each will develop different interests. The idea here is to allow these interests to develop freely and be used within the framework of the family farm. A child with computer skills can develop customer lists or be in charge of Internet research for appropriate projects. The artist can illustrate brochures and newsletters. The mechanic can keep all the machinery going and learn welding.

This is one good reason to have a diversified farm, including the marketing diversity. By creating all these different elements, each child's talents can be expressed and utilized within the operation.

Since we operate an apprenticeship program, we have even more opportunities to have our sanity tested with all the different interests these guys bring to us. Of course, I like them to take advantage of my extensive agriculture book collection. I've been collecting these over the years and have some real gems--especially old ones.

We once had an apprentice who wanted to make a buckskin outfit. I mustered all the charity I could find and said okay. He went after it with a passion. He called a local slaughterhouse and procured a bunch of hides. Then he went to the woods for a day and cut a load of saplings for stretching frames. We could scarcely get in and out of the shop. Using baler twine as stretching string, he had the equipment shed completely blocked off with skins on these frames.

Numerous experiments finally yielded a workable scraper made out of a scrap piece of automobile leaf spring and he set to work with a vengeance scraping those hides. It was cold and the hides worked better warm, so he took our heat lamps that we use in the chick brooders and set them up, working into the wee hours of the morning. I can't remember all the procedure, but I know somewhere along the line he went back to the slaughterhouse and brought home a small bucket of deer brains, which Teresa dutifully slurried in the kitchen blender. The brain tanning went well and soon it was time to break the hides.

Because it was cold, he brought the skins in one at a time and set by the fire downstairs, breaking them over a stand he'd made. When he left at the end of his year, he sure enough had a pile of buckskin. There were times during this process that I felt like he would benefit far more by reading some of these books I cherish. Notice, these are books that I cherish.

He'd been through college and was tired of books. He desperately needed some hands on gratification. He'd been wanting to do this since childhood. I have a sneaking suspicion that the buckskin project was far more useful to him than reading my books. The question I had to ask myself was this: "Do I have enough confidence in him to trust him to pursue his dreams?"

That really is the ultimate question. I think most of us adults do not trust ourselves, and in turn, do not trust our kids. This ubiquitous lack of assurance expresses itself in our trying to define every move of the next generation.

We must provide a reassuring environment that gives the children opportunities to fail. Without the chance to fail, they will never succeed. In order to eliminate failure, we deny success. You can't have it both ways. You can't hedge in your children that they can never fail without also hedging them so much that they can never experience the elation of personal success.

I know situations where 30-year-old men can't touch a 2x4 on the lumber pile until they ask Dad. That is a travesty. Early on, make a conscious decision to give the children some rein. Let them take a scrap board you may think could be useful and drill holes in it, pound nails in it, and tear it up. The benefits gained from creating an environment of freedom to explore, learn and achieve on the part of the children will be far more valuable than a two-bit board. Boards can be replaced; children can't.

Early on, Daniel became acquainted with Tom Brown and John McPherson, who have written numerous books about survival skills and living off the land. This was really never an interest of mine, but I have been fascinated with some of the things Daniel has done. He built deadfalls and figure four traps, made cordage out of deer sinew, and has now graduated to making the absolutely best and most consistent deer tenderloin jerky in a wood cookstove of any I've ever eaten.

It was worth the wait. Was any of this a waste of time? Some people would probably say so. But I vehemently disagree. He graduated. Yes, he now has the skills, and what's more, the love, to do some incredible things that directly enhance the farm and our quality of life. Preserving the passion must be one of the greatest gifts parents can give children. Society enjoys beating down personal interests, pounding on the nail that sticks out a little farther than the rest.

One of our chief jobs as parents is to keep the excitement alive. And a critical factor in doing that is giving the children freedom to explore their interests. If we do that, they will find ways to make it meaningful to the family enterprise. That's the payoff.

Chapter 11

Commandment 4
Create Investment Opportunities

Let the children invest in the farm. I never cease to be amazed at the number of farmers that act like they can take their assets into eternity.

Everyone knows you can't take it with you. Children understand that their parents aren't going to be around forever, but parents seem to act as if they'll be immortal. And the truth is that when children don't feel like they can get a piece of the pie now, what's the use in trying to hope for it later?

I know middle-aged children, farming with parents and grandparents, even, who can't even buy a tractor tire. The surest way to affirm the next generation's control is to allow the children to own portions early. The earlier the better.

Investment always stimulates long-term commitment. As

soon as Dad knew I wanted to farm, he quit buying things. If we needed a new piece of equipment, I bought it. If we needed a building permit for a building, I went down and got it in my name.

This did several things. First, it gave me a clear understanding that there would be something here for me when I wanted to take over. There was never a nebulous realm out there, a sort of never, never land, in which my input and future here was in question. It was clear that my place was assured and there would be room for me.

Too many times as children begin growing into the farm or family business, their physical equity is delayed for too long. This lack of equity makes them question not only their future there, but also whether or not the parents really want them there. The delay even makes them question their real contribution, their real value, to the operation. As soon as someone puts an ownership stake into a component of the business, it changes the emotional relationship to the enterprise.

It's no longer "theirs," it's "ours"; and beyond that, "let me take you over here and show you my part." Maybe it's just a tire, but the ownership element works its magic within the recesses of the young spirit, and creates a satisfaction impossible to acquire until that investment occurs.

This is the beginning of the next generation gaining par with the previous one. Most normal children want an heirloom from the family home. But the opposite is also true. Children love to see their stick-men drawings and blue ribbons hang on the refrigerator door. What parent doesn't proudly display children's artwork on the desk or on the wall in the bedroom? This is as important for the children as it is for the parents.

Secondly, investing gave me a clearer understanding of just how much things cost. It's easy to complain about not having a new door on the barn or a fancy cordless power tool. But as soon as the child begins purchasing these things, reality comes home fast. Suddenly the children develop a great appreciation for the financial shrewdness of their parents.

How many times does a parent retort to a child: "Well, if you had to buy it, maybe you wouldn't have one either." Interchanges of this sort are a natural and common occurrence between parents and children. The children think the parents are miserly tightwads who don't care about progressive infrastructure and the parents think the children can never be satisfied with anything. Butting heads, the relationship falters and the parents and children eventually part ways. Mom and Dad are left to plod along and age on the farm while the children seek their fortune elsewhere. If for nothing else than the economic education offered by the investment, parents need to promote it. Their future is at stake.

Third, investment makes the new generation more careful about how the infrastructure gets treated. Horsing around on the machinery lessens when children know that they will pay for the replacement. When a piece of barn roof begins flapping in the wind, it suddenly becomes noticed by the child. Before, it was just the old barn starting to fall apart. Stupid barn. Why did they build it that way anyway?

But when its replacement carries a $12,000 price tag, the care and attention it deserves jumps astronomically. Keeping the children in the dark and denying them opportunities to invest is a dead-end economically and emotionally. The goal is for the children to realize that the parents' things are the children's things. Why do so many farms get stuck on this point? It seems like most family businesses stumble right here. The graceful balance between the older generation letting go and the younger generation coming on board, and all the delicate elements that must come together in harmony are incredibly difficult to maintain. Most of the time we're emphasizing one side more than the other.

I've heard middle-aged children say time and again: "Dad just treats me like a hired hand." In fact, I knew a young man once whose father told him he could hire any migrant worker and replace the son. Talk about cutting. Do you know how that must have made the son feel? Within a year, he took off.

I realize that the infrastructure on the farm represents a lifetime of care and investment on the parents' part, but at some point it needs to physically and emotionally change ownership. Encouraging the children to invest in the new and replacement infrastructure will keep those kids vested in the farm. It's their vested interest that is at stake, and parents hold the key to creation or denial.

Certainly a parent giving this latitude risks seeing the child take an investment beating. Perhaps the idea just doesn't work out like the child thought it would. Better to learn the lesson now while Dad and Mom can help pick up the pieces, than wager the whole shebang on the idea after the safety net is gone. Plenty of multi-generational family farms have been lost because after the death of the parents the child who stayed on the farm decided to take all that equity and build a confinement poultry house.

Of course, after reading this book you'll understand at the get-go why a confinement poultry house would be an unwise investment. And if this book keeps one family from making that devastating blunder, the whole effort will have been worth it. I received a call a couple of years ago from an environmental group in West Virginia asking me to come and do a couple of seminars for all the poultry farmers there who were losing their multi-generational farms because of the economic reality within the poultry industry.

Tragically, these farmers owed a lending institution the money, and it didn't matter how unprofitable running the house became, the lender still wanted its money. So the farmers had to keep running chickens through the poultry houses, even if it paid only $1 per hour for their labor, just to generate the income to keep the lenders at bay. Of course, the worst part of this story is that often these houses are built with the full blessing of the elderly parents. The whole family views it as the only way to stay in farming.

Children need to appreciate that parental wisdom does matter. But the only way the parents can keep the children coming for advice is to allow the children investment opportunities that will allow them to spread their own wings. Mark Twain commented about how when he was 18 his Dad didn't seem to have a lick of sense. But by the time he was 23, it was amazing how much his Dad had learned. Things haven't changed much since Mark Twain.

Stephen Covey deals with this issue when he articulates the principle of emotional equity. The way the parents gain an emotional bank account is by letting the children invest in the business. That investment opens up a respect and openness on the part of the children toward the parents, and creates room for parents to advise their children without appearing tyrannical or arbitrary.

How can parents ever know if the children can be entrusted

with the assets of the estate if they do not watch what the children do with what they acquire? If the children are good stewards, persevere, invest wisely and do a good job, the parents can rest assured that the farm will be in good hands. Without this track record, the parents will go to their grave wondering what in the world will happen to the farm when they pass away.

Settling that stewardship ability earlier rather than later can save a lot of mistrust and assumptions in the critical last years. The way many children handle their assets, if I were a parent I wouldn't leave them a dime. I'd be out beating the bushes for a responsible young person passionate about farming before I left it to a couple of children who took no interest in the farm except what it was worth at market value. Farming at best takes all the grit and determination that can be dredged up within the human spirit. Anyone can lose a farm; keeping it takes a rare person indeed.

The only way children can be trusted with an inheritance is to let them invest and establish a reputation of success before the inheritance actually occurs. As the investment accumulates, the ownership feeling gradually shifts over, switching the child-to-parents dependency to parent-to-child. In other words, as the parents age, the farm equity gradually shifts over to the children. This makes the parents dependent on the children rather than the other way around.

This dependency keeps crotchety old men from throwing their weight around and practicing emotional blackmail on their children and grandchildren. Mark it down. When you see a farmer unwilling to release control to the children, you are seeing the beginnings of a complaining, accusing, unappreciative old man. The patterns for attitudes late in life are cultivated early. And nothing reveals more about where our heart lies than how we view our assets.

Unwillingness or inability to relinquish the reins shows a heart that has no appreciation for the next generation, no trust, and no faith. Children feel this denial of investment opportunity acutely. Dad just wants to protect what's his, I know. But that doesn't do the next generation any good. Instead of standing on a launching pad, the next generation stands in shackles. And people in shackles yearn to be free. Freedom comes from leaving the farm.

This need not occur. Letting the children invest, encouraging them to invest, and patiently watching the stewardship of those investments will help the kids to love the farm.

Commandment 4: Create Investment Opportunities

Chapter 12

Commandment 5 Encourage Separate Child Business

Encourage the children to have their own business separate from Mom and Dad's enterprises. Everyone needs to have a personal claim to fame. No one wants to go through life in the shadow of someone else. We all yearn for something we can point to and say: "I did that."

Children raised in a healthy, familial environment are extremely entrepreneurial and natural marketers. Who can turn down the pleading eyes of a child? That's why Girl Scout cookies and candy bar fund-raisers work so well. Children naturally wiggle into your heart and get a positive response when any other marketer would probably fail.

Too often parents let a slice of the family business suffice for the child's autonomous element, and that is not the same. Having a herd of cows with three designated as the child's is not the same as the child having a completely separate enterprise. A subset, or simply having a designated portion of the existing enterprise does not offer personal ownership and control.

The separate enterprise principle was certainly instrumental in my own growing up years. My brother Art, three years my senior, decided to raise rabbits. He built pens and we visited an old codger in a nearby town who had quite a rabbit enterprise. He acted as Art's mentor and got him going.

When I was 10, I wanted something to call my own. I had duly won my Snoopy dog for selling a minimum number of magazine subscriptions in some school fund-raiser. But that didn't offer what a business would. I had a great-uncle who at the time was one of Ohio's premier egg producers. My paternal grandmother was one of four girls and a boy. Uncle Omer was the boy, and he had quite an operation.

As a child, I was enthralled when we went to visit him because he was a full-time farmer. In those days poultry confinement houses were just coming into vogue and he had both range layers and brand new double-decker confinement houses. This was way before the caged permutation. Even in these houses, the birds had bedding and were what we call today "loose housed." The birds layed in hay-bedded nest boxes and he had a fellow working for him who gathered the eggs. When this guy wanted a snack during his workday, he would just throw his head back and crack an egg above his mouth, swallowing the egg in one dramatic gulp.

Uncle Omer's range poultry arrangement was amazing. He had literally 10,000 layers out on range. Predators were one of his biggest problems so he had two strands of electric wire strung at three inches and about 10 inches off the ground. He plugged these into straight 110 or 220, I can't remember which. If a raccoon or 'possum hit that wire, it would just fry him dead on the spot.

Uncle Omer's business card impressed me even as a child. It was the most treasured possession in my wallet when I began carrying one. I proudly displayed that card to school friends. It was a picture of his range chickens spelling his name on a hillside. The way they did it was to let the birds get hungry for a day and then they fed them shelled corn in the configuration of letters spelling his name. A picture from the opposing hill yielded a perfectly clear presentation of O.M. SMITH. That card finally composted in my wallet, but it was a real treasure.

I'm explaining this story to make the point that business ideas, or fancies, in children come from many different sources. I'm sure no adult in my life at the time had any idea what impact that card and those couple of visits to Uncle Omer's farm did to me. Our lives are shaped by circumstances that only come into focus years later. Your children and mine are touched daily with words, pictures and experiences that trigger fantasies and help shape life goals. These experiences prick the subconscious into creating pictures of "what I'd like to do." Often we aren't even aware of what's happening at the time. Only in retrospect can we appreciate the power of the experience.

Equally strong were the visits to my paternal grandfather's half-acre small fruit and vegetable garden in Indiana. The entire garden was bounded by an overhead grape arbor that to me was the most Edenic thing I'd ever seen. I happen to love grapes, and these purple, succulent Concords awakened every sensual pleasure normal to a little child. The yellow bees buzzing about the dripping juice, the sweet aroma of ripened fruit and the sight of all that bounty oozed into the depths of my spirit.

His garden, laid out in neat squares, had grass pathways defining the neatly cultivated plots. I could go out and visit the garden without fear of dragging mud into the house. His strawberries, carefully limited to five daughter stems apiece, bore the biggest, juiciest berries I had ever seen. Raspberries, carefully pruned and held up by wire cylinders, could actually be picked without fighting thorns and a tangled mess. Order prevailed. And abundance. It was everywhere. With the maturity of adulthood, I now can appreciate Grandpa's desire to farm full-time but settling for part-time.

At the time, I elevated Uncle Omer and Grandpa to divinity because their production and abundance seemed almost beyond reach for us. But Dad, after the Venezuela experience, had to start over from scratch. It takes time to put together a model that looks like a going concern. And so it was natural that gardening and poultry would appear to me to have the most potential for a going concern. This at a time when we were chopping thistles, Dad and Mom were both working in town to finish paying off the farm, and our own eroded hillsides, gradually greening up with the trees we'd planted, still seemed a long way behind what I sensed at Grandpa's and Uncle Omer's.

At 10 years old, I bought 50 as-hatched mixed heavy breed chickens from Sears and Roebuck. We put a cardboard box on the workshop bench in the basement and attached a heat lamp to keep them warm. These yielded 18 pullets and 32 roosters, a typical split I've seen repeated countless times. I've never been able to get hatchery folks to fess up to the "as hatched" myth, but what I don't know about the world of hatcheries would fill a library.

We had an old chicken house outside the back yard. I fenced it in and began my layer enterprise. They did extremely well and I began selling eggs to neighbors and folks at church. I'd put a few dozen in the front basket of my bicycle and pedal to the neighbors, delivering my eggs. Dad showed me how to scald and eviscerate the roosters and I was off and running.

The main point of this whole story is this: Dad didn't know anything about chickens. I was the resident expert. And that made all the difference. When people visited and began asking about the chickens, Dad would come and get me to show them the poultry operation.

By the same token, if I had problems, he couldn't solve them for me. Instead, he became a co-learner, a co-conspirator in the great struggle between profit and loss. If I had a problem, I had to find the solution myself. A couple of years later I decided I needed more efficient processing equipment. I placed a classified ad in a poultry magazine, which yielded two phone calls, both from places several hours away.

We hitched up a trailer to our 1963 Chrysler Newport and headed out, before daylight, one morning to see what these guys had. The first place had two really nice units, a scalder and one-at-a-time picker, stainless steel, but the $250 price tag was way more than I could stand. I remember having to walk away from those shiny units as if it were yesterday. I had the desire to have them, but knew they were out of my financial reach. The willingness to admit that and walk away was one of the most powerful lessons of my early years.

Then we went way out on the Eastern Shore of Virginia and visited the other man, who was about 75 years old and had given up processing his chickens when he retired. He had a much older galvanized setup for $75. I jumped on it. We loaded up the whole kit and caboodle in the trailer and headed home. By this time it was beginning to get dark and when Dad turned the lights on, the alternator in the car went out.

That was in the days when you could still run a good distance without the alternator, but it was dark. We got on the highway and Dad ran with just parking lights for an hour until we found a shop that was open. I can remember the ensuing tenseness of the situation, but all turned out well and we made it home. That little feather picker still works fine and has been loaned out to any number of friends and neighbors over the years. I have never forgotten the encouragement and sacrifice that trip represented. It was one of my first "business trips."

At about 12 or 13 years old, I wanted to expand sales so we checked into the Saturday morning Curb Market in Staunton. It was a post-depression institution, started to give cash-starved farmers a city sales venue. They had never had a child sell there, so the powers that be decided that instead of my having to join an Extension Homemaker's Club (the market operated under the blessings of the Extension Service) I could just join 4-H instead and that would suffice to show my bonafide desire to abide by the establishment's rules and regulations.

Throughout my teen years, I was up every Saturday morning at 4:00 a.m. to sell at the Curb Market, which opened at 6:00 a.m. and ran until about 11:00 a.m. Eventually, it had dwindled to two elderly matrons. One sold baked goods and made divine potato salad, which awakened within me an insatiable love for homemade sweet pickles. In fact, I did not taste anything as divine until the first time I merited a Christmas family shindig at Teresa's house and found--you guessed it--homemade sweet pickles.

To this day I joke that I made my decision then and there to marry this girl whose family preserved the 14-day homemade sweet pickle secret in dramatically sensuous fashion. Was that confusing? Did you misunderstand? I meant that the pickles were sensuous. I said sensuous, not sensual. Okay, enough of this.

The other lady came from a diversified homestead and sold a potpourri of cured pork, vegetables and baked goods. Our booth added my eggs and vegetables, Art's rabbits, Dad's homemade yogurt from our couple of Guernsey cows that we hand-milked, as well as butter and cottage cheese and beef. I wouldn't take a college education for what I learned at the Curb Market. The word "organic" was scarcely invented and the clientele was evenly split between folks who came for a bargain by buying directly from the farmer, and the ones who came for high quality by buying directly from the farmer.

In retrospect, I'm sure I added my own spice to the market, and probably gave these two grandmas plenty of stories. They taught me the gentle balance between coddling a customer and saying "take it or leave it." Pricing, arranging my wares, listening to what people wanted. These were invaluable business skills I still use every day.

My fresh egg business continued to expand through my high school years until I supplied two small restaurants, two public school cafeterias and on Saturday mornings I'd sell 60 dozen eggs before 8:00 a.m. During this time, my involvement in 4-H expanded and I began taking advantage of the many projects and contests it offered. I was a state winner twice and competed nationally a couple of times. Two highlights stand out. The first was being picked by a consortium of poultry science specialists working on 4-H poultry projects, which were funded at the time by Kentucky Fried Chicken. These specialists wanted a 4-Her to attend their final meeting to get a firsthand critique of what they had planned. The Virginia extension poultry science specialist told the committee members that he had just the guy.

We drove out to Louisville, Kentucky and I hobnobbed with these movers and shakers for a couple of days. But the big deal occurred the second evening when we all drove over to eat at The Colonel's Lady, the restaurant at Colonel Sanders' house. It was all family style service and about halfway through the meal, Colonel Sanders walked in. Of course, these adults I was with were KFC executives and personal friends with the Colonel. I was the only young person in the group.

Colonel Sanders sat right down next to me and for the next two hours entertained us with stories and his own magnetic personality. It remains one of the most memorable events of my youth.

The second powerful experience was attending national 4-H congress which at that time was held every year in Chicago's Conrad Hilton. When I won the trip, Mom contacted Paul Harvey to see if I could get an audience. He wrote back a kind and hospitable letter to have me come on over and see him when I got into town. The second day of the week-long gala I hiked the half dozen blocks down Michigan Avenue to the ABC studios.

The secretary greeted me warmly and said Paul was finishing up his broadcast and would see me shortly. You must understand that in our politically and news-savvy family, Paul Harvey was an icon. Without a television in the home, entertainment centered around books, radio and the record player. In the mid-1960s, during the height of the Cold War, Paul Harvey recorded a record titled The Uncommon Man, which in typical Harvey style saluted the person who dared to rise above mediocrity. In a day when our culture was beginning to kowtow to the average, and shining excellence was painted as greed and selfishness, Paul Harvey dared people to escape the mundane and strive for the stars.

I literally memorized this record as a teen, and can still recite large portions of it verbatim. In addition, Harvey narrated the between-segment portions of another record we played often, titled Yesterday's Voices, which had actual recordings of great preachers, beginning with Dwight L. Moody and ending with Peter Marshall. Harvey's salutatory "Good Morning Americans" and his signatory "Good day" were daily features in our house on his broadcasts.

So here I was, sitting at his secretary's desk in Chicago, a teenage poultry farmer from Virginia's obscure hinterland village, Swoope. In only a few moments, he strode into the office, asking for me by name. We shook hands warmly and he invited me into his office, where he finished typing what would be his Saturday noon broadcast. A staunch Seventh Day Adventist, Harvey prerecords his Saturday broadcasts to preserve the Sabbath. I'll never forget his first question: "Who is Joel Salatin?"

I didn't have an answer. The question took me so aback I still have no idea what I said, but whatever it was he laughed good naturedly and handed me an autographed copy of the first edition of his son's brand new brainstorm, The Rest of the Story. Then we went into the recording studio and chatted while the sound engineers readied the equipment on the other side of the glass. He then began about 60 seconds of voice staccatos and exercises to limber up his mouth, then the red light came on and we were hot.

After he recorded the broadcast, we chatted a few minutes before he finally excused himself to go back to work. I walked back to the convention as high as a kite. All of these special experiences came directly because of my chicken business, which only gained notoriety because it was really mine. All of these experiences went into building my self concept and my direction in life, and they were truly positive experiences.

Now fast forward a dozen years. Daniel is eight years old and has grown up listening to stories about his uncle Art's rabbits. The old pens, hanging in the barn rafters where we stored them when Art shut down the project to go to college, beckon Daniel to do something with rabbits. Why? Who knows? This is just the magic and mystery of entrepreneurship. You never know what will trigger the opportunity, or trigger the desire. As parents, we just need to be sensitive to these little nudges coming from the kids and then release the children to pursue these ambitions.

About that time a friend called looking for a home for about three rabbits. A move to the city required a new home for the pets. Daniel took the rabbits, lowered Art's old rabbit pens and went into business. He joined 4-H and competed in judging, presentation and public speaking contests, amassing offices, ribbons and notoriety. At 19, he was elected president of Virginia 4-H.

He has spoken at sustainable agriculture conferences around the country about his rabbit enterprise and conducted countless tours of his facilities. He won his trip to national 4-H congress as well, putting together lifetime experiences that will make him who he is. Since I don't know anything about rabbits, he has had to learn it all the hard way. Whenever I see tables of books somewhere, I always look for something about rabbits. He has quite a personal collection now of rabbit books and is becoming an expert in the whole business.

He continues to confront problems because of the non-medicated, forage-based model he's pursuing. One of the largest producers--and certainly the youngest--in the mid-Atlantic region, he does all his record keeping, supplies purchasing, pen building and marketing. He has already helped many smaller children with their fledgling enterprises with his custom-built pens and breeding stock. He is learning the principles of business.

When he was about 12 years old, he was buying rabbit feed down at the feed store. The clerk tried to get him to buy a whole ton instead of just a half ton. Partly teasing, the clerk said: "Why don't you just go ahead and buy a whole ton?"

"Because I don't have enough money for that," Daniel laughed.

The guy looked up at me and said: "That's a lesson most adults haven't learned." It reminded me of that shiny chicken picker when I was the same age. Folks, this is the stuff of life. This is as good as it gets. This is value transferral, legacies and family.

Rachel is our artsy, craftsy, kitchen wizard. Teresa and I can't draw a stick-man, but Rachel, even as a toddler, drew beautiful flowers and could illustrate stories. She can go for a walk and pick a handful of wildflowers, arranging them into a perfect bouquet. When I'm traveling doing farm conferences, I always look at the book tables for something about flowers or baking for her to add to her collection.

She knits, crochets, sews, bakes, flower gardens and has her own enterprises. She cleans a lady's house in town twice a month. Certainly we have things around here she can do, but this was an opportunity that fell into her lap and she wanted to do it, for herself. It's important to give the kids freedom to know that they can pursue their ideas. We don't need to concern ourselves with the fact that this is probably not a career for her. The important thing at this stage in her life is to support her go-get-'im attitude. She wants to pursue this so we need to give her some rein.

In addition, she has quite a following among our farm clientele for her zucchini bread. Daniel grows her zucchinis and she turns them into the best bread anywhere. When customers exclaim over her, and little grandmothers bend down and pinch her cheeks--she just loves that--and effuse: "Oh, your zucchini bread was the best treat any of the garden club ladies have had in a year. Everybody carried on over it so." What do you think that does to the self image of a child?

Rachel absolutely radiates after these encounters. One of the deepest needs of the human heart is to be needed. "I need you" is a powerful phrase, and awakens all kinds of latent energy. To be needed, to know we're fulfilling a person's life, is a wonderfully affirming experience.

When Rachel was about 10 years old, the rest of us got busy and tilled up a section of pasture for a flower garden. She poured over her books, talked to adult gardening friends and gradually populated it with her own specimens. It is beautiful, and she tends it, makes refinements in it, and owns it, both figuratively and objectively. She is now in charge of all landscaping around the farm. In fact, a couple of years ago we completely changed the position of all the brooder houses after she developed a plan that was much more aesthetic than my hodgepodge.

The enterprise can be simple or complex. When both children were extremely small, they made potholders on those little loop looms, selling them to folks at church for Christmas gifts. They went through a period of time making refrigerator magnets with little beads on a peg mold. As the children age, they can take on different or larger enterprises.

Working for others can be a complementary angle to this whole issue. Rachel's house cleaning is an example of an ancillary moneymaker that allows her to be the boss--in Mom's house, she just can't be the boss--and receive compliments from someone besides her own family. This is essential in building self worth and a can-do spirit.

Daniel helped a friend plant trees and another friend erect log homes from time to time. The paycheck was not nearly as useful as the experience of taking orders from someone besides Dad, and their subsequent praise of his work.

For teaching thrift, accounting, marketing, communication, perseverence, responsibility, diligence, courage and all the characteristics parents would like to cultivate in their children, business has no parallel. When the child is the boss and the business success or failure is entirely up to him, all these wonderful characteristics are cultivated in the life of the children. The tears over the failures and the exultation of the successes can be shared multi-generationally, but when the youngster owns the tears and joy, the experience gains a priceless richness. And the experiences endear the youngsters to the oldsters in a mutually respectful bond strong enough to last a lifetime.

Commandment 5: Encourage Separate Child Business

Chapter 13

Commandment 6
Maintain Humor

Keep a healthy sense of humor. This may seem trite, but I can tell you this admonition needs to be shouted from the rooftops in farm country. The need for humor is a little unique to farming--and one big reason why many farm kids leave for the city as soon as they are old enough to assert independence.

This may seem contrived to you, especially if you aren't familiar with farmers. But I guarantee this is a typical farmer exchange.

Farmer Brown: "Howdy, Jones, how's it goin'?
Farmer Jones: "Ohhhh, can't complain, Brown. What's this weather gonna do?"
Farmer Brown: "Looks like it's gonna rain."
Farmer Jones: "Yeah, it does. But the weatherman says it's gonna go north and we'll just see the clouds."
Farmer Brown: "Well, you can't trust the weatherman. He don't know nothin'."

Farmer Jones: "My knee's actin' up. That usually means low pressure."

Farmer Brown: "Oh, my back aches somethin' terrible. But that don't mean nothin', 'cause it aches all the time."

Farmer Jones: "We need the rain, but I'm tryin' to get this one piece of hay in so I can turn it and plant some corn."

Farmers Brown: "That piece worn out already?"

Farmer Jones: "Yeah, weeds and brush. Can't keep anything anymore."

Farmer Brown: "I thought you sprayed it last year."

Farmer Jones: "Yeah, might as well have seeded the weeds. Chemicals don't work, but I don't know what else to do."

Farmer Brown: "It'd be good if the rain would hold off and then come on right after you plant the corn."

Farmer Jones: "Oh, it'll probably rain on my hay and then turn dry as soon as I plant the corn."

Farmer Brown: (Laughing) "Naw, it'll probably drop a gully washer and send all the corn down the creek."

Farmer Jones: "Might as well. Corn ain't worth nothin' anyway."

Farmer Brown: "Ain't that the truth. Bringin' less now than when I was a boy. All them infernal foreigners. They'll ruin us yet."

Farmer Jones: "Can't do nothin' about it. Government's tryin' to put us out of business as fast as it can. They might as well come on and take it all. That's what they want, dadburn 'em."

Farmer Brown: "Looks like it's clearin' a bit. That hay might be about ready."

Farmer Jones: "Yeah, probably is, but I can't make it 'til my son gets off work."

Farmer Brown: "I know how that is. Can't get no help."
Farmer Jones: "Young people, they don't hang around no more. Don't know what's got into them. Got to have the latest car, the fanciest house. I just don't know."
Farmer Brown: "You don't see mine around either, do ya'? Can't blame 'em. Long hours, dirt, dust, smell. Can't make a dime."
Farmer Jones: "They can leave if they want to, but one day they'll wake up and not find any food on the store shelves. What then?"
Farmer Brown: "Then it'll be our turn, and we'll drive the prices up and make 'em pay, buddy. Make 'em pay."
Farmer Jones: "Serve 'em right."
Farmer Brown: "Good to see ya'. I'd better get to the house before the missus jumps on me for bein' late to dinner (all farmers call lunch "dinner").
Farmer Jones: "See ya' in a couple of days if I'm still kickin'."

Folks, that is not made up. You can hear a conversation just like that a hundred times a day across the fruited plain. Remember that the average age of the American farmer is approaching 60. These are lonely old couples whose children left 20 years ago for greener pickings in the city. They watch their Walkman-entranced grandchildren visit, the multinational corporations increase in power and government policies continue to give bigger subsidies to everybody but them.

They watch the rich and famous on TV and feel like life has passed them by. They are very much like Patsy Clairmont's son who explained why he didn't want to go to school: "It's too hard, too long, and too boring."

To which she replied: "You've just described life. Now get on the bus."

This world of ill-dealt hands is the world farm children grow up in. And they don't remember the good times until the golden shackles of their Dilbert cubicle fills them with anguish. Then they remember the simpler life. And they want to go back. But it's really a memory trick, whitewashed over the years to reveal only the pleasantries. The reason I know this is true is because if it were as enticing as they now recall it, they would have never left in the first place. True, some people will leave regardless. But generally the fault lies in the parents for not creating emotional and economic opportunities for the next generation.

Take time to enjoy the humor on the farm. Watch the calves run across the field, kicking up their heels. Tussle with your children in the hay mow after the last bale is in. Enjoy the sunset.

Go on picnics, even if it's just to the back field. Too often farmers feel like they have to go elsewhere for their enjoyment. Build a pond with an adjacent barbecue pit. Create your own recreation and entertainment. In a diversified farm, who needs to go away for excitement? After watching the tree you're cutting down fall the wrong way, into the fork of a nearby tree, and cutting it down, only to have it fall into the fork of another tree, getting underneath to cut the third tree to let the whole shebang fall at once is plenty of excitement to get your heart ticking.

Anyone who ever played with a bunch of pigs will have experience comparable to the best slides in a water park. Pigs love to be rubbed under the bellies. I've rubbed many a hog until, in sheer ecstatic contentment, it just falls over on its side, grunting its quiet sweet nothings.

A fellow in Minnesota told me about his city cousins who used to visit the farm to go "hog fishin'." The kids would go out by the woods and cut a couple of long saplings for rods, tie on about eight feet of baler twine, and then tie an ear of corn on the end for bait. Then they'd go to the hog pen and dangle that ear of corn out over the pigs. Of course, the curious pigs would come over and sniff around on that ear of corn. Finally one would grab the ear in its mouth and begin tugging. At this point, the child would scream: "Look! I've got a big 'un!" A tug-of-war would develop between child and pig until one finally let go. It was great fun, and I'm sure could be resurrected today as a money-making agri-recreation package.

No child can handle push, push, push all the time. Parents must be willing to kick back and enjoy the atmosphere the farm offers. Lots of people would pay big money for what we see out our window any day of the week. The symmetry of the plants and the acrobatics of the animals yield countless opportunities to laugh. I may not make the world's highest salary, but I sure have a great office. Dwelling on the low prices, the fickle weather and the weeds will drain the drama right out of the farm experience.

The emotional and financial pressures on farmers are immense. They are real, and I appreciate that. But that just means we need to exert extra effort to combat the temptations to, as we say, "Cry in our beer." Part of the annual "to do" list needs to include items that will enhance quality of life. Certainly sometimes going away can be healthy and create a great memory. We've certainly enjoyed our family getaways.

But as animal behaviorist Bud Williams says: "If you have to get away to get your batteries recharged, don't come back." His point is that if our vocation does not energize us, then we're probably doing the wrong thing or at least doing it the wrong way. If our vocation wears us down and puts bags under our eyes, something is way out of whack. And unfortunately, that is the experience of most farmers and the attitude most farmers portray to their children.

I've often thought it would make great house decorations to hang a picture in each room of animals doing things appropriate to that room. A prominent shot of pigs at a trough could grace the dining room wall. A shot of the vegetable harvest could adorn the kitchen. A cow properly hunched and dumping a healthy cowpie would be perfect for the bathroom wall opposite the toilet. And a picture of the cats piled up sleeping on the back porch would be wonderful for the bedroom. Maybe the living room could feature a flock of laying hens clucking away. For some reason, Teresa hasn't okayed these ideas yet.

Turning work into games and contests is all part of this humor. Reading aloud as a family is all part of this. Probably the biggest element is just stopping long enough to enjoy the surroundings with your child.

As adults, and especially as farmers, the work is never done. We can fill our days from dawn to dusk with important, meaningful work. We do not punch a time clock. A 40-hour work week does not exist. We literally must force ourselves to take the time to laugh with our kids. It doesn't take an hour a day. A little bit goes a long way. This time is leveraged probably more heavily than any other time segment in the day.

Children naturally fear their parents and hold them in awe. Dad is strong and can do anything. Mom is organized and can heal any wound. And these feelings are as they should be. But in addition, we as parents must covet a "best friend" attitude with our kids, and humor plays a vital role in accomplishing that. We can incur fear without friendship, but what kind of relationship is that? It's one that ends when the kids can get away.

If we want the children to stay with us, take our baton and continue running around the same track, we must cultivate laughter together. It will endear us to our children in remarkable ways. What we're after is a relationship in which the kids can't wait to bring their friends over and introduce them to us. When that begins to happen, the future is bright.

Commandment 6: Maintain Humor

Chapter 14

Commandment 7
Pay the Children

Pay the children for their labor, making sure they understand that the farm generated the income. They need to understand the source of the wealth. That way they'll develop gratitude in the right direction.

I don't believe in giving children an allowance. For those of you already doing this, hear me out on this one. I know allowances are supposed to be a way to give children money so they can learn proper stewardship. I'm all for kids having money and learning proper stewardship. But money should be earned, not guaranteed.

No person should receive pay just for breathing. This teaches that society owes me something. Society doesn't owe you or me anything except the right to life, liberty and the pursuit of property. You did know that when Thomas Jefferson originally penned those words, he didn't write "pursuit of happiness." This will come as another shock to folks who think the best way to steward the land is to give it all to the government.

No, the founders viewed the freedom to own and utilize property as an inalienable right, a right, in fact, uniquely American. That was a critical aspect that separated the American experiment from other governments of the day. Land in Europe belonged to the various kings, and was doled out according to despotic whim. The idea of any commoner actually being able to own land was a new idea, and the founders so equated it with being able to pursue dreams that they just edited it to "happiness," never dreaming that anyone would interpret that fundamental right as anything except property. My, how things change.

I think allowances teach children that society owes them a living, and that is a terrible thing to foster on a child. The principle is translated societally into minimum wage laws, which disemploy countless of entry level people and create a me-sided employee mentality. Suddenly what I earn is not up to me, it's up to some Sugar Daddy in Washington to decree how much I should be guaranteed for darkening the door of my employer's business.

This kind of wrong-headed thinking creates an egocentric worker who takes no responsibility for where he is on the pay scale. Thinking that the employer owes him one, he tends to slack on the job, pinch some paper clips or ballpoint pens when he can, and blame everyone but himself when he's passed over for a promotion. The sooner our children nail down the reality that they earn respect, they earn trust, they earn responsibility and they earn financial rewards, the sooner they will grow up to be happy, well-adjusted, productive citizens.

Allowances do not stimulate children to go above and beyond the call of duty. They are like pass-fail school courses, which do not stimulate excellence. If all you get is what you get, why put out for more? Do what's necessary to keep from being yelled at, or to keep from failing, and that meets everyone's expectation, including your own.

Since I've dug myself in this deep, I might as well throw on the tombstone and say I'm not even in favor of wages or salaries. I realize that anyone can do a good job and receive promotions, but this is often too delayed for the average laborer to understand. I like bonuses, commissions and performance contracts whereby a person has limitless earning potential. The potential is tied directly to performance, and not to a supervisor's political and subjective whims.

Too many people have been denied advancement because they were the wrong religion, color or ethnic background when promotions are based on the subjective evaluation of a supervisor. Some folks get passed over because a supervisor is afraid the promotion may mean he, the supervisor, gets passed up.

I had a friend in college who worked in an automobile plant in Indiana during the summers. His job was to push some steel plates through a machine that bent them into the proper shape. After spending about a half hour learning how to operate the machine, he finished the pile stacked there in just four hours. He asked the supervisor for more material.

"Oh, that is all that machine is allowed to do in a day," the supervisor informed him. Of course, it was a union plant.

"But, tomorrow I'll be able to do that pile in less than three hours," my friend complained. He had been taught to give a full day's work for a full day's pay. He had been taught that there was no free lunch.

"Sorry," said the supervisor. "That machine is under quota.

The operator can only do that many pieces."

"Well, when I finish, can I leave?" queried my enterprising friend. After all, if he could work half a day and get full day's pay, he could go do another job and double his income.

"No, you have to stay at the station the full eight hours, in case we get audited or something. Rules say you have to be at the workstation, but you don't have to be feeding the machine."

So my friend spent the summer reading in the automobile plant. I think it was about this time that the Japanese took over the American automobile market. I certainly hope this kind of thing isn't happening today, but it's the kind of mentality that develops when pay is not tied directly, and I mean directly, to performance.

Allowances create a welfare mentality, a "you owe me because I'm your kid" kind of thinking. This matures into "you owe me because I'm working for you." Permutations on this theme abound in our culture. "You owe me a ribbon because I entered the contest" is one, as well as "you owe me a passing grade because I showed up for class." Forget it. Unfortunately, our culture has fallen into this gimme, gimme trap and created a subculture of dependents.

I heard a most fascinating radio talk show one night. The guest was an author who said, and I kid you not, that we should do away with child support payments in divorce settlements. My ears pricked up and I leaned forward to hear what this lunatic would say next. This was wild stuff. It turned out he was opposed to spousal support of any kind, including the laws that gave authorities the go-ahead to track down deadbeat dads and make them pay.

His reasoning was nothing short of fascinating, and I'll

confess I don't agree or disagree. I haven't sorted it all out. I just think it's one of those things that shows us just how little we know about things and how easily we get sucked in to the popular paradigm without questioning its philosophical foundations. He said that it was this guaranteed security net that allowed women to trash their homes and their husbands and not be submissive, loving, domestic wives.

He said men are inherently wild and part of the woman's job is to tame his wildness. Yes, I can see that. When the force of the government can make him take care of her, regardless of marital status, then it offers her license to not work at the marriage, to not tame the wild man. She and the children get an allowance either way. In essence, then, these laws, while not necessarily encouraging neglect of spousal responsibilities in marriage, certainly make it easy for headstrong women to abandon their conjugal duties. He assured the audience that if we did away with these post-divorce safety nets for women, they would work at staying married and taming their husbands, using the talents that women possess, to affectionately domesticate their wild men.

Isn't that amazing? Like I said, I haven't just thought it all through, but it is a fascinating take. Which brings us back to all this allowance and security issue. Guaranteed annual incomes, guaranteed minimum wage, guaranteed this, that and the other never bring out the best in people. They always push recipients to a lower level of performance, all the while creating a dependency and a "you owe me one" world view.

The final swipe at allowances: we never appreciate what we don't earn. This is one reason why I dislike government grants. I realize filling out paperwork could be considered earning something, but the amount of money doled out for this project and that which should be financed by private initiative and grit is obscene. And a farmer whose paycheck comes from a government grant will never be as creative at problem solving, and will never appreciate what he does rightfully earn, as the farmer who built his operation up through free enterprise.

I cringe at the carrots the government holds out in front of farmers. From being paid not to farm to being paid to build visitor chalets, there is no end to the sugar plums offered at taxpayer expense. And I watch the sustainable agriculture community practically fawn over this "seed" money coming from Washington, watch people grovel for it like pigs at a trough. This money comes with strings attached and oversight that compromises the integrity of the project. The farmer works for a committee, or the government, not himself.

If you built your farm business on the back of the American taxpayer, when you look back in 20 years you'll never appreciate it like the farmer who scraped and scratched to get it going. We've created dependency classes in virtually every facet of our culture. The things that private business, private creativity were supposed to solve and did solve for a couple of centuries, building the most beloved, benevolent nation (notwithstanding what we did to the South and to Native Americans) in history are now done at the behest of politicians and bureaucrats.

Perhaps the best example of this is what happened to Malabar Farm, Louis Bromfield's home in Ohio, after he died. He willed it to the state of Ohio to maintain as a bastion of innovative eco-friendly models. During his tenure, the farm thrived and touched millions of lives around the world because one man, Bromfield, stayed at the controls and poured his sweat, money and vision into it. As soon as the government took it over, it fell into disrepair, chemical usage and oblivion. Seldom has something as stellar fallen as quickly. Why anyone with his apparent common sense would think that the government would appreciate what he'd done or keep it up is beyond me.

I love his books. I love his techniques. I love what he did. He was a leader for all time. But he did not end well. And that is unfortunate. The state couldn't appreciate what he'd done when they received something for nothing any more than he would have been as successful if the state had given him a million dollars and said: "Go see what you can do." He would not have scraped, pushed a sharp pencil, and come to the conclusions he did, I guarantee it. He arrived at his solutions because it was his money. And when you spend your own money, it makes you a little more interested in real solutions, not just solutions that will satisfy the existing paradigm.

The best grand example of this is the money given to support land grant colleges, which have lost sight of their direction and been bought off by corporations to rubber stamp chemical research and genetic engineering. The colleges couldn't appreciate their public trust because they weren't earning the taxpayer's money that was fueling them. So it was a simple matter to be taken over by vested interests, and to spurn their charters.

I heard a speaker say that no governmental institution besides the United State Department of Agriculture has worked so hard to annihilate its constituency. Usually government agencies grow their constituency bases. The education department tries to build more schools, enroll more students and hire more teachers. The defense folks try to build more bombers and recruit more soldiers. The welfare folks try to sign up more people and raise the poverty floor to push more people into the welfare rolls.

But the USDA destroys its constituency systematically and efficiently. What if the size of the USDA were tied to the net income of farmers? That seems like a fair standard to me. When the net income goes up, the budget goes up. When the net income goes down, obviously the farmer's source of information is failing, and the budget should go down. Forget the allowances. Forget the guarantees. Let's get back to some performance-based budgets and put some accountability back into the system.

By now, you understand that I have a real problem with allowances. But let's talk a little about the positive. What constitutes good remuneration? Children should be paid for performance.

That doesn't mean they should get paid for everything. They need to learn that team players do some things just because they are part of the team. For the most part, daily household chores should not be financially rewarded. Those are things that need to be done just to function efficiently. Completing school assignments, taking a bath and keeping a clean room are basic requirements of being human. They're equivalent to sweeping the house and fixing dinner.

But children should be paid for their labor. They should not be taken advantage of by being made slaves. Discuss the big jobs with them, including remuneration. Where possible, tie the pay directly to performance.

My nephew for a couple of years became a turkey subcontractor for us. He took the turkeys when they came out of the brooder and grew them for $2.50 per bird. He created his own schedule and was in complete control of them. Of course, we'd pop in a couple of times during the growout just to make sure things looked good. But because his pay was linked to dressed birds, he was meticulous about their care. He rarely lost one.

If a child is gathering eggs, tie his pay to the number of

salable eggs. Don't tie it to total eggs gathered, because he may not exercise proper caution to bring them to the processing room uncracked. Furthermore, eggs that are too dirty or stained to be cleaned don't count. That encourages meticulous nest box bedding, and timely gathering. All pay should be like this, linked to the actual income derived from the product.

If you have an orchard, picking apples can be tied to bushels picked. This encourages pickers to not drop fruit. Perhaps if dropping seems too large, take a walk and agree on what looks about right. If too many are on the ground, deduct something. Don't let slipshod work go uncorrected. The ideal is to so clearly articulate the job and its performance-based pay gradients, that no one could feel cheated when the money finally gets squared away. Both parties need a clear understanding of all the variables in the equation. If the expectations and rewards are clearly understood by both parties, it minimizes the chances for being miffed about fairness.

Now the big question: How much? Obviously, I can't tell you how much to pay your children. But let me lay out a basic principle--give them what it would have taken you in today's dollars to feel so good about the work that you'd want to do it again. How's that for an answer?

Stingy farmers are ubiquitous. For some reason, farmers feel like they are grossly underpaid, underappreciated and cheated by society. As a class, we farmers feel as tromped on as probably any vocational group in society. I don't think aspiring singers and artists feel as downtrodden as today's farmer. This creates in farmers a miserly disposition, and a close-to-the-chest financial mentality.

A young friend finishing up an animal science degree at a prominent land grant university was telling me about the economics of using anabolic steroids in cattle. I told him that it seemed unfair because these hormones simply make the cells expand, take on water, and thereby weigh more. When the consumer goes to the meat counter, she's not paying more for nutrition; she's just buying more water.

"Well, the consumer's been screwing the farmer all these years, so I figure it's about time to get even," he responded. Isn't that a wonderful lesson for aspiring young farmers to learn? Sounds like the ag professors in that university were inculcating all the finest objectives in their graduates.

I know adult children working with their parents on the farm who have to ask for a dollar to get a soft drink at the local filling station. If an adult child is working hard and doing a conscientious job, contributing to the welfare of the farm, he should be rewarded in kind.

As soon as Dad knew I wanted the farm, he drastically reduced the money he put into it, as well as the income he derived from it. Unfortunately, many parents don't appreciate their children until, in frustration, the children leave and then the parents must fend for themselves. Parents, deprive yourselves of some monetary advantage to encourage the youngster to persevere and stick with it.

I realize that often the income is tight. But I would rather deny myself, as the owner, an income if things are tight, and make sure my children get it, than hold onto it myself and make them feel like the farm doesn't pay anything. The question is: Where will the greatest long-term marginal reaction occur? Will the greatest long-term payback be Mom and Dad putting more money in their CD at the bank, or will it be in making sure the children understand how much they are appreciated?

As soon as the farm is a going concern, the money needs to be appropriated according to performance. Adolescent sons and daughters do a pile of work around farms, and should be paid accordingly. I mean thousands of dollars, not just the $600 allowable as non-reported casual labor. Money is certainly not the only incentive, but it is a powerful one. "Show me the money" is not just a trite saying. It is a prime motivator.

The ability to earn $10,000 a year as an adolescent is a tremendous incentive to a young buck--or doe. These kids will work their heart out for you when they see that kind of earning potential. Remember, they are rubbing shoulders all the time with people who plan to earn a lot more than that in city jobs. The way to head them off at the pass is to show them the potential early on, so the stars in their eyes are focused on the family business, not somebody else's promises.

I know many farmers would respond to me that such a pay rate would mean the kids would get more than the parents. What's wrong with that? Once the children shoulder the work load, why in the world do the parents have to amass a bigger estate? We'll talk more about this issue in the chapter on inheritance, but I think for now it behooves us to understand the dynamics of the situation. If I continue squirreling away money for a rainy day, meanwhile depriving my child of a real sense that this farm thing can support him, I'll need all that money and more as I age.

But if I forget about my retirement fund and funnel that to the kids, they will have the team spirit and the appreciation to compound that interest in communal family function worth more economically and emotionally--not to mention spiritually--than all the interest the CD will produce.

Now don't take me to an extreme here. I'm not saying that

middle aged farming parents should not put any money away. There is a balance in all of this. I'm coming out swinging on this issue because I think it is one of the most overlooked issues in the agriculture community. For some reason, the average farm couple expects their adult children to slave away at the farm business for most of their lives, as paupers, until suddenly one day the parents die and kids receive an inheritance windfall.

I'm not suggesting we give the kids money they don't deserve, but I am begging farmers to offer their children the same respect, financially, that they would offer a hired man. They deserve at least that much.

This whole issue dovetails nicely with the separate enterprise issue. With good pay coming from the core farm business, and with a no-ceiling personal enterprise, your children will see earning potential on the farm that will make them understand it as a viable vocational option. The kind of pay going to most farm children wouldn't be enough to entice a baboon to do it for a living. This mentality creates brain drain in the countryside as the best and brightest go to the cities and the D students stay home on the farm. The poor dumb redneck farmer is stereotypical for a reason.

I implore you, farmers, be gracious to your kids. Deprive yourself, if need be. Don't just toss them a crumb from time to time. Give them the firstfruits. Give them the very best. They'll spot your heart's desire and respond accordingly. Everyone will be happier.

Chapter 15

Commandment 8
Praise, Praise, Praise

Praise your children.

This is another command that seems obvious enough, but in my experience it is neglected more and more as the children get older. For some reason, infants receive oodles of positive reinforcement, but then when they don't keep their innocence and begin to disobey and give looks of defiance, parents begin conveniently forgetting how important it is to stroke.

When the children do a good job, the parents say nothing because after all, that was the minimum requirement. Obedience is expected, not rewarded. The normal routine is that good work receives no special notice, and the parents and children rock along in relatively silent co-existence.

But when the child bends over one nail, or misses one garden weed, then the parents fuss and fuss. Many children leave the farm because whatever they did was never quite "good enough." Are you a fussy parent? Did you have fussy parents?

Of all the areas I touch when speaking about family friendly farming, I'd say this is the one that hits more people in the deepest recesses of their spirit than any other single topic. I've had men come up to me at the end of a conference presentation, tears rolling down their cheeks, and say: "I'm 52 years old. I could never please my Dad. I am just now getting to the point where I'm ready to try something."

I watch the audience, and routinely I see wives cast sidelong glances at their husbands, who are sitting stoically, unflinching. Then I watch big crocodile tears slowly descend the wife's soft cheeks and she snuggles up closer to her husband, who is now trying to fight back an emotional response. Although this issue plagues both men and women, I think more often it's a man problem.

After doing my family friendly farming presentation at a Minnesota conference, a friend gave me this report: "I was sitting behind a father-son duo. The dad was probably 80 and the son about 50. Halfway through, the father leaned over and told the son this speaker wasn't making any sense. To which the son replied: 'This is exactly what you need to hear.' "

We men are driven to excel, to win, to compete. We don't go shopping, we go hunting. We don't go out to dinner for romance, we go for food. We don't go on vacation to create a memory, we go to improve our golf game . . . and go to as many interesting places as possible. Men and women are different--beautifully different.

For you men reading this, if you're not sure about your status on the fussy index, ask your wives. They know when you're being too hard on the kids, when the push, push, push has become a little too intense.

Men tend to not gush about all their plans, so the children often don't know that Dad's eye was on that board for a special project 10 years hence. All Junior knew was that it was lying there and looked like a pretty good board for a bicycle jump. Dad, how are you going to respond to that? I'm here to tell you that on the continuum of life, the damage you'll cause by fussing is a whole lot more than the damage of losing a two-bit board.

Dad and son are working on a project, nailing boards up. Everything's going along fine--and quiet--when suddenly Junior bends a nail over. Dad's fussing breaks the silence, and Junior dies inside, thinking: "I can never do anything right."

As an apprentice pattern maker in a Chrysler automobile plant, my Dad honed his woodworking skills until he could carpenter to a fault. He could build furniture. He even made his own gouges and wood chisels. His woodworking toolbox was full of calipers and sophisticated measuring tools, special little bubble levels and planes. From my earliest memories I have looked longingly into that box and wished I had the creativity and ability to use those tools. Each handmade wood gouge, carefully wrapped in a protective sheath, could produce the most magnificent wooden objects under Dad's careful craftsmanship.

What about me? Probably the easiest way to describe my carpentry skills is to say that my tape measurers are marked off in inches. That's right, just inches. Whenever other people come over to work on projects with me, they bring these special tapes that have a bunch of lines between the inches. I have no idea what they represent. They call back and forth measurements that end in "quarter" and "eighth." I'm standing there completely bewildered saying: "Quarter what? Eighth what?"

I haven't been able to find any of those special tapes yet, but I'll bet someday I might find where they get those things. Maybe I could learn how to count off all those little marks someday. So far, I've gotten along pretty well with my inches. And what's more, Dad let me get by with my inches.

That's the punch line in this story. After the laughs about my dysfunctional carpentry ability, you must realize that Dad never once complained about my 85 degree angles. Not once. Do you know how much restraint, how much charity that took on his part? The older I get, the more I appreciate what that did for me.

There are actually people in this country who think I'm a pretty sharp designer. You ought to see folks out there measuring my chicken shelters--and they're using those special tapes. Why are they 2 ft. x 10 ft. by 12 ft.? Because inches is all my tapes say. But do you think for a moment that I would have been emotionally free to take the initiative and build these shelters, or the host of other construction projects I've done, had Dad fussed about my unsquare corners? Not on your life.

And that's what the 52-year-old was saying, his eyes flushed with tears. His Dad tied him up emotionally by fussing at everything he did. That infamous rural expression "It won't work" has destroyed the spirit of many a child. How many children, after constructing their first birdhouse, proudly display it to Dad, and hear him grunt: "Top's a little crooked."

You may as well hit your child in the face with a 2x4. I've actually been present when these exchanges take place, and I want to wring that Dad's neck. I want to shake him and scream: "Do you know what you just did to the spirit of that boy?"

But the average Dad would just respond: "He needs to know that if it's worth doing, it's worth doing well. We promote excellence, not shoddy work."

"But Dad," I implore: "Don't you understand that if it's worth doing, it's worth doing poorly at least. And if you destroy his zest for doing at all, he'll never become skilled enough to do beautiful work."

My heart breaks for these children whose complaining, fussy Dads--and occasionally Moms--assess their children's work with a critical eye FIRST. I'm not suggesting that we never offer constructive criticism. But we earn the right to criticize only after we've invested enough positive reinforcement in the child's "emotional bank account," to use Stephen Covey's metaphor.

A child composes a story. Cute little story. "Look, I wrote a story, I wrote a story," Junior gushes.

"Hmmm," says Mom or Dad. "You have five misspelled words." And the child's face falls. The spirit is broken. How many more stories will Junior risk before giving up? Why is it that we can go to a little league baseball game and banter the game away with two hours full of "'Atta boy. Swing batter. You got 'im" but when we come home it's complain, fuss and criticize?

I'm middle aged. Dad passed away 12 years ago. I still have to choke back the tears sometimes when I finish a project, realizing how much I miss his affirmation. In fact, I'm choking tears back right now. That's how powerful a child's desire is to please his parents, and how much a child needs to know his parents are proud of him. These needs are woven by God into the foundational fibers of the human spirit, and nothing can change them. We will forever be built with the need to hear "good job" from our parents. And when it is begrudgingly offered, or given not at all, a great abyss forms in the depth of our spirit. We are timid to try anything, lest we fail. We certainly don't feel good about ourselves, so we in turn take that out on our kids.

Since my youth, I have studied farm families whose children all left for town. In a four-child family, the odds of keeping them all on the farm are fairly remote, and perhaps not even a goal worth pursuing. But at least one child of two, three or four should catch a passion for the farm. When none wants to come back, I have found that the most common denominator is this issue of fussy Dads. My point is that the overwhelming majority of farms pass out of the family not because the children didn't want it, but because one or both parents drove the children away.

The burden for creating a place where the children want to stay rests not on the children, but on the parents. I'm sure there are exceptions to this rule, but I'd bet 90 percent or better of the breakup cases are due to the parents driving the children away. When nothing I do is good enough, I'm going to head somewhere where people appreciate my contribution.

I remember well the first tailgate I made for a hay wagon. The two uprights were really good but my angled cross pieces on the inside were way out of whack. It really looked quite awful. But I worked hard on it and finally got it done. Dad came out to inspect it, and I feared I may have crossed the line this time. I could even tell it was messed up without my inch tape measure. He just looked at it and laughed: "Well, it works. That's all that matters."

Do you know how liberating that was? Oh, how my spirit soared. In effect, he said to me: "I give you permission to try anything you want to. Whatever your initiative finds to do, go for it. And I'll be there and be proud of whatever you accomplish."

I actually build pretty good ones now.

Oh, parents, praise your children. The first time Daniel drove the tractor all by himself, I told everybody at church the next Sunday what a great driver he was. He stood there holding my hand, just beaming. Praise your children to others. Praise them for completing jobs, sticking to it, staying with you until the last hay bale went in the mow. Don't just come out of the barn together and sigh: "Well, at least that's over for the day."

Rather, look your son straight in the eye and tell him how much you appreciated his help, and praise him for how hard he worked. Thank him for being such a good son, and ask him to help you be a deserving father. Heap that praise on your children and overlook the little things that can make you fuss.

Raymond and Dorothy Moore, homeschooling gurus and authors of half a dozen books on the subject, say that the stimulus to make children learn is adult-child interaction. In other words, the interaction is probably more important than the details of the lesson. That is simple, yet profound. Other studies show that the average adult-child interaction averages 17 minutes per day in a public school classroom.

Looking back, I think this is the greatest contribution homeschooling combined with self-employment has made in our family--praise time. I've been there for the first word that the kids read. I've been there for the first nail Daniel pounded in, the first flower that bloomed in Rachel's garden. Parents who send their children off to daycare or who work away from the home give over to some other adult the opportunities to share these praise opportunities with their children.

I wouldn't take a million dollars for the praise opportunities I've had with my kids over the past twenty years. Not only have Teresa and I been there for the firsts, but we've had many more opportunities to praise. When the family is separated for a large portion of the day, all the affirmation must be crammed into a couple short hours. All of us know that those hours are typically wind-down, exhaustion hours in most households.

Who has the energy to get pumped up about anything, except to rise off the sofa and scream: "Keep it down back there!" Fuss, fuss, fuss. But in the cottage industry, home schooled paradigm, the day is not only filled with adult-child interaction and all the positive stimuli that generates, but also hundreds of opportunities to praise, to affirm.

Daniel has competed in 4-H presentation and public speaking contests since he was old enough to enter. Teresa, Rachel and I have been to virtually every competition. When he became old enough to compete on the state level, we traveled to Virginia Tech to watch him perform. He won every time he competed on the state level. Then we'd go back and watch him receive his pendant. Seems like any parent would do this, doesn't it?

What has become more and more disconcerting to me is that when I sit there in the competition room, it is devoid of other parents. Last year was the icing on the cake for me. The pinnacle of speaking competition is the state 4-H public speaking contest. Daniel had qualified a couple of years before, but decided to go with his illustrated talk instead of the speech. His final eligible year, he decided to go for the big one.

He won the county, then district, and the big state showdown came. A field of 30 or so contestants competed for the public speaking award. He had never had so much competition. You would think the room would be full of siblings and parents who came to cheer their kids on, to see their charges compete in the premier contest of 4-H congress. Teresa, Rachel and I had left home at 5:00 a.m. to be sure and get there in time. We noticed about two other parents in the room. That was it.

I was appalled. Here was the cream of the crop, the best of the best, and their parents could not even take a day off from work to share the thrill of the moment. I was first angry, then sad. No wonder families are dysfunctional. No wonder kids turn to drugs, peers and non-familial vocations for approval and affirmation. Is it the young peoples' fault that Dad and Mom aren't in the room? The burden for keeping it together rests squarely on the shoulders of Mom and Dad, not the kids. They'll respond just like all kids respond to praise, warmth and unconditional love. If you don't know what's gotten into your kids, the answer is probably nothing--primarily you and your affirmation. If you don't get into them, something else will, and it might not be pretty.

Oh, of course, Daniel won it. And we were there to cry and cheer and share the moment, hugs all around, and proud Dad, Mom and sister to big brother. Big strapping 19-year-old, celebrating with his family. Didn't matter that friends were all around. Didn't matter that he was away from home. He was a family teen, and proud of it. We couldn't have been prouder. Yes, that's our Daniel. Why would parents give up these opportunities? Why?

Parents, get with it. The time is short. If you think those little kiddos are going to wait around for you to get involved with their life, think again. You're going to turn around tomorrow and they're going to be as big as you are. You won't be able to just "make them" do things. You will be drawing on your emotional investments made throughout the formative years. And you'll need every deposit you can find.

The cottage industry homeschooling model offers countless opportunities throughout the day to affirm your children. Just for the strength of that alone, I would like to see thousands of families pursue this option. It might be financially tough, but it will yield dividends in the arena that matters most.

When your child washes dishes, overlook the little egg stain on plate number five. Really, it's not a big deal. It'll come off sooner or later. If it takes your children an hour to wash dishes, something's messed up. Their zest for the work has been destroyed somewhere along the line. Either you have not infused them with a joy for the work or a critical spirit has dampened their zeal. How many times have children brought home a report card with all A's and one B, only to hear about the one B?

Why do parents do this to their kids? I'm not a psychologist, but I know these interchanges are real, and have lasting impacts. I suppose much of this comes from being brought up that way, or being loved conditionally. One of the big problems with this issue is that I'm convinced most of us have blinders on when it comes to self-evaluation.

That is why I ask husbands to listen to their wives. When she says: "Honey, back off. You're being too hard on him," don't just dismiss it as nagging. Fellows, these ladies are trying to help us in this area of great weakness. This is why the Bible says to dwell with our wives "according to knowledge." Most of us guys are so busy with our projects and pounding our chests with great accomplishments that we're clueless about what our wives are feeling. They are blessed with that wonderful nurturing intuition. We need to learn from them.

Let's commit ourselves to praising often, praising long, and praising hard. Put some gusto in it. The kids will brighten and thrive, and will want to stick around us back-patting parents for a long time.

Commandment 8: Praise, Praise, Praise

Chapter 16

Commandment 9
Enjoy Your Vocation

Enjoy the farm yourself.

Complaining, bitter, griping parents don't raise children who love the farm. Kids gravitate to the things their parents are passionate about and flee things their parents despise. Of course, the "enduring the job" syndrome is not unique to farming, but farmers certainly have a healthy dose of it.

How many farm kids have heard their parents say: "I want a better life for you?" By "better life" they mean easier work, shorter hours, paid vacations, paid medical, 401(k) and social standing in the community. The sad truth of farming is that the models most farmers use create burnout and depression. I've watched farmers with gung ho kids ready to tackle the farm and really think they can make it pay, be put off by the parents because "if it didn't work for me, it won't work for you." Most farmers labor under the paradigm that "you can't make any money farmin'."

They surround themselves with other depressed farmers who stoke the boiler of "woe is me." The weather will always be worse, the prices always lower, the plants and animals always more sick, and city people always more stupid. The outlook is bleak.

Most farmers personify the moonshiners on the long-running TV show Hee Haw, crying:

> *"Gloom, despair, and agony on me.*
> *Deep dark depression, excessive misery.*
> *If it weren't for bad luck, I'd have no luck at all.*
> *Gloom, despair, and agony on me."*

Country music thrives on the bad luck theme. You've heard the old joke about what you get when you play a country music song backwards?--you get your wife back, your dog back, your job back, your truck back . . . I'd better quit now before you country music lovers begin throwing tomatoes.

But it's true that the country persona carries with it this black cloud of bad luck and disaster. For too long farmers have not really made progress, and have not had a positive attitude to solve their problems. One of my favorite passages in all the old farming books I've ever read comes from a little book by C.C. Bowsfield published in 1913 titled <u>Making the Farm Pay</u>:

> *"The average land owner, or the old-fashioned farmer, as he is sometimes referred to, has a great deal of practical knowledge, and yet is deficient in some of the most salient requirements. He may know how to produce a good crop and not know how to sell it to the best advantage. No citizen surpasses him in the skill and industry with which he performs his labor, but in many cases his time is frittered*

away with the least profitable of products, while he overlooks opportunities to meet a constant market demand for articles which return large profits.

Worse than this, he follows a method which turns agricultural work into drudgery, and his sons and daughters forsake the farm home as soon as they are old enough to assert a little independence. At this point the greatest failures are to be recorded. A situation has developed as a result of these existing conditions in the country which is a serious menace to American society. The farmers are deprived of the earnest, intelligent help which naturally belongs to them, rural society loses one of its best elements, the cities are overcrowded and all parties at interest are losers. The nation itself is injured.

Farm life need not be more irksome than clerking or running a typewriter. It ought to be made much more attractive and it can also be vastly more profitable than it is. Better homes and more social enjoyment, with greater contentment and happiness, will come to dwellers in the country when they grasp the eternal truth that they have the noblest vocation on earth and one that may be made to yield an income fully as large as that of the average city business man."

Isn't that amazing? Even though that passage is nearly 100 years old, it could have been written today. That's the thing about truth--it never changes. Now I would like to pose a question, with all due respect, to you older farmers reading this: "Do you believe?"

That really is a fundamental question, because it speaks directly to the idea of enjoying the farm. If you do not believe a farm can yield an enjoyable living, it never will. You don't just happen to fall into enjoyment. You create your enjoyment.

Bowsfield's insights can scarcely be bettered. Indulge another passage from the same book:

> *"It ought to be the aim of every farmer to accomplish these definite results:*
>
> *Increase profits by enlarging production at a fixed expense.*
> *Diversify crops and all other profits so as to distribute labor evenly throughout the year.*
> *Secure a regular income at all seasons by supplying customers with poultry and dairy products, vegetables, beef, pork, etc.*
> *Shorten the work-day to ten hours, provide a comfortable home, improve the appearance of the premises and try to make life enjoyable.*
> *Let the young people have a little money from the production of fruit, flowers, vegetables and experimental crops.*
> *Teach them to plan work for themselves and to love the country."*

Is that good or what? What an astute man. And yet can you name for me even one of these being attained by our current industrial agriculture paradigm? Every single piece of advice from the USDA is exactly opposite these. Perhaps the first one is debatable, about increased production at fixed costs. But it's debatable because I haven't seen much about fixed costs. All I've seen is increased production. That's why farmers can't produce themselves out of their debt.

Not one of these goals fits the current conventional agricultural paradigm. You'd think the establishment could hit at least one or two. But it can't even hit one. The sustainable agriculture groups that are starting to pay attention to these ideas, of course, are finally getting a little sop from the agriculture establishment. But it's only a sop. The establishment is still running pell mell toward monocultures, mechanization, cloning, hybridization and depopulating the countryside.

Dad was always looking for new ways, better ways. He always said the ultimate goal of a job was to work yourself out of it. Part of the excitement I remember growing up was the challenge of progressing. Each year I saw the hay get heavier and thicker, the cattle better, the work more efficient. Unfortunately, most agricultural models require more veterinarian care, more pharmaceuticals, more chemicals and more dangerous heavy metal in a death spiral to economic and social ruin.

A farmer out in Missouri told me he was going to shut down his confinement hog operation. This was a typical state-of-the-art farrow-to-finish operation on slatted floors with a huge manure lagoon underneath. These slats were made of a heavy plastic material.

I asked him why he was shutting it down, and he said it was because he realized that his first conscious thought, every morning when he woke up, was wondering what sow fell through the slats into the manure lagoon this time. He said that was no way to live, and I agree. At the time, his eldest daughter was planning to leave the farm. Within two years of shutting down the confinement facility and turning his attention to more sensible enterprises, the daughter saw such a big change in her Dad's countenance and attitude that she decided she wanted to come back and work alongside him. Isn't that a fairytale ending?

Commandment 9: Enjoy Your Vocation

Unfortunately I can't convince many farmers that there really can be an enjoyable future in farming. That notion threatens the average beat-down farm codger because it means he spent his life doing the wrong thing. Nobody wants to believe they threw away their life. As a result, these bitter old farmers are virtually unreachable. My heart aches for all the acreage under their control, being grossly underutilized. That land could capture far more sunshine and sequester tons of carbon. But the pasture gets thinner, the weeds thicker, and the barns more unkempt.

You can't argue these guys into a different outlook. As Allan Nation, editor of Stockman Grass Farmer magazine says: "It's easier to change the person than change the person's mind." Unfortunately, much of this 60-year-old farmer generation will have to die off before enjoyable paradigms can be put in place on any kind of massive scale.

In spite of this reality, I still plead for two things. First, that farmers institute enjoyable models, and second, that even if you can't enjoy it, at least let the youngsters who believe they can, have at it. Don't deny these youngsters the chance to build a different future.

This is one reason I promote alternative paradigms. If we are to salvage our farm children, we must change the way we farm. We must institute models that are emotionally, environmentally and economically enhancing enough to romance the next generation into farming. And if our models do not do that, the problem is in the model, not agriculture.

Children will naturally be drawn to our passion. Enjoying our farm vocation is one of the keys to creating that magnetic draw that will keep the children near.

Chapter 17

Commandment 10
Back Off from Personal Domains

Parents, back off. That means you. Quit looking around. I'm talking to you. I'm talking to all the meddling, micro-managing parents out there who can't understand that if you don't remove the old mulch, new seedlings can never get started.

If the groundwork of the other commandments has been properly established, giving your children wiggle room should not be risky. The reason parents can't give wiggle room, for the most part, is because they haven't created an environment that allows the children to establish their credibility. Without credibility, trust cannot occur, and without trust, freedom is a pipedream.

One of the things people in our community marveled about when Dad passed away 13 years ago was how smoothly the farm weathered the transition. By the time of his death, Dad was working for me, not me for him. I don't want that to sound irreverent, but it illustrated perhaps his greatest single trait. He really loved to see his students successful.

In his accounting business, his greatest joy (besides reducing a client's taxes) was to teach a skill to a bookkeeper, or a farmer, that would reduce his accounting time. When a client took an idea and reduced Dad's time in half, that just made Dad's day. He could then take that time and teach someone else. Nothing pleased him more on the farm than to watch me read the books on his shelf, to learn from him, and to implement those ideas. Indeed, he loved to back off, to let his children get the glory. I have come more and more to appreciate not only how difficult that is, but how important it is to making kids love the farm.

Part of this is simply the skill of delegation. Good leaders know how to delegate. Insecurity, in both the boss and the underling, creates the dynamics leading to a meddlesome relationship. I know parents who live in mortal fear of their children. "What will they do next?" is the question nagging at their subconscious.

Instead of a quiet, steadfast knowledge of their children's solid values, these parents are always in a state of anxiety about tomorrow's shocking discovery. When the children are extremely small, we as parents are the instruments God uses to shape their value system. Once the children are bigger than we are, if they aren't coming to us for advice, we're not going to help the situation by smothering them with it unsolicited.

I know this is the temptation because being older and wiser we can see the result of the direction they are headed, and we want desperately to have a different outcome. But once the children are in their late teens and early 20s, our coercion options narrow significantly. Anyone who grows plants or animals knows the value of a good start. Those first few days in the brooder set the stage for the entire life's performance of a broiler chick. In the garden, a good seedbed, proper moisture and temperature, are critical for the performance of any vegetable.

If you plant green beans just before a week of cold, nasty weather, those stunted plants will never perform as well as the ones planted two weeks afterward when all the conditions are perfect. And that is why I belabored some of the earlier points in this section, because as we come to the final commandment--backing off--we must ponder all the earlier ones, and how they were implemented in the formative years.

I see struggling young couples all the time opting for day care with their little ones, and my heart breaks for these families. They will fight ear infections, runny noses, additional bad habits and a divided loyalty. I've watched other parents make the decision to downscale their living standards, put the socio-economic climb on hold, and invest that time and energy in their children. In a short ten years, you can see clearly which family opted for which lifestyle. I would like to say religion plays the key factor here, but in my experience, it has virtually nothing to do with the parent-child bond and trust. New Age cosmic worshipers, Moslems, Christians--if the infants and parents spend time, the chances for happy, balanced, responsible children increase astronomically.

Gradually backing off gives the children the freedom to explore, and gives the parents accountability opportunities. If we don't begin backing off early, we can never test our progress. In his wonderful book You Can't Fire Me, I'm Your Father! Neil Koenig has some fascinating things to say about the values that bring out the best in people. He has this to say about two of them:

"Responsibility

This is the number-one motivator. Nothing motivates people like responsibility. People, whether little or big, are much more likely to excel when they are given responsibility for something from beginning to end. This is in contrast to simple obedience, where people, whether little or big, take no ownership for carrying through: 'it's not my responsibility. I was just doing what I was told.'

Accountability

This has to do with consequences--positive consequences for good performance (thank-yous, recognition, praise and tangible rewards), negative consequences for poor performers, non-performers and subtracting performers (the hot seat of explanation and learning for improvement, correction, coaching, goal setting, withholding of recognition and reward or, when all else fails, dismissal). In thirty years of experience working with families, I have concluded that accountability is the least practiced value of all. Parents, executive leaders, managers and supervisors are simply reluctant to mete out consequences, either positive or negative."

How can we establish accountability without offering responsibility to our children? How can we offer responsibility without backing off? The two go hand in hand. Looked at another way, how will the children feel a sense of accomplishment and gain in self confidence if we don't back off?

By the way, Koenig offers other values that he has found bring out the best in people. They are:

The Golden Rule
Honesty

Trustworthiness
Respect
Knowledge/learning
Work Ethic
Cooperation/Independence
Excellence (not perfectionism)
Kindness/Compassion
Optimism
Money as Morally Neutral
Forgiveness
Fun/Laughter/Play

Koenig's book is the best I've ever read on the elements of successful family businesses. Others offer advice on trusts, foundations and financial contracts, but this one deals with the relationships, the real dirt-under-the-fingernails stuff.

And as we come to the conclusion of this entire section about making your children like the farm, I'd like to quote Koenig's section titled "Getting Permission to Succeed." I had prepared these 10 Commandments several years ago, and when I ran across this freshly-published section during the flurry of cramming that always precedes writing a book, I was amazed, and gratified, to see how closely our lists agreed.

"There are truly blessed people who get permission to succeed at home from an early age. Their parents do such things as:

- **take genuine interest** in their children's interests;
- **listen carefully** to their children's thoughts, feelings, imaginations and dreams;
- **applaud** their achievements;
- **read to their children**, and listen to them read;

- **take a keen and active interest** in their school experience;
- **expose their children** to life's incredible variety;
- **give age-appropriate responsibility** and decision-making opportunities;
- **set high but reasonable** and age-appropriate standards and expectations;
- **correct mistakes** without hurting feelings;
- **give plenty of latitude** for trying things;
- **show understanding** and forgiveness for mistakes;
- **stick to appropriate consequences** for irresponsibility
- **set an example of optimism about the future**; and
- **speak well of work** and its possibilities and rewards."

Isn't that a wonderful list? If all families could do these in correct balance, we'd live in a different world. None of us gets them all just right. But we can certainly strive for the balance, and we can appreciate our own shortcomings if we contemplate these ideas. Then we can take corrective action, putting attention on our areas of weakness.

Romancing the Next Generation

Chapter 18

Pleasant Farms: Aesthetic and Aromatic

A family friendly farm is both aesthetically and aromatically pleasing.

This makes it a fun place to live and a fun place to show friends. Anybody can show off a home. But more and more, those who want to show off their farm are harder to find.

The two senses of sight and smell are the two most critical for diagnosing disease. How can we tell if a wound is infected? We smell the foul odor. A visibly swollen joint lets us know about sprains. In the kitchen, sniffing what's in the refrigerator lets us know if the food has outlived its usefulness.

The one I can never figure out is the way moms notice their kids' facial color. I don't know how many times Teresa has asked: "Don't you think Rachel's color is a little off?" I can never tell. The kids always look healthy to me. It's definitely a motherly attribute.

Listen to their conversations talking about their children, and invariably you'll hear them matching color stories. We men just sail off into oblivion. The kids could have the plague for all I know, but as long as they're tagging along and breathing, I don't have a clue whether they're pinkish, reddish, whitish or greenish.

Be that as it may, we do use sight and smell to distinguish health. The first sign that food is bad is a foul smell, or taste, which is really just an extension of smell. Generally we don't just dismiss the foul odor or the funny looking stuff as being unimportant. We ask for second opinions. The jar of old green beans with a half-inch of moss growing on the top makes its rounds with appropriate groans before we pitch it to the chickens.

Mark it down. From food production to food processing, and that includes farms, a foul odor or unsightliness indicates a wrong model. Period. End of discussion, amen.

Many states have passed "Right to Farm" laws to protect unsightly fecal factory concentration camp farmers from litigious neighbors or environmental groups. The laws should be renamed "Right to Stink up the Neighborhood and Pollute the Water" laws. They are meant to protect obnoxious farmers from neighbors' redress. And they have been enacted because foul odor and unsightliness characterize the modern industrial farm.

Half the stuff we use on farms, which are supposed to be bastions of life, end in the Latin suffix "-cide" which means death. Herbicide, pesticide, grubicide, infanticide, genocide--they're all kind of the same. Have you ever smelled that stuff? The label reads: "Don't breathe it, don't get it on your clothes, dispose empty container in accordance with hazardous materials." Farmers apply thousands of tons of that stuff to food and think nothing of it.

But let us get beyond the food. How about employees who have been killed by inhaling a little too much of these concoctions? For the sake of worker safety alone, no one should use these things. Notice that I did not say they should be outlawed. As opposed as I am to deadly chemicals, I do not believe the government should intervene in their use. Of course, if the government were not involved in their use, neither would it be involved in their promotion through government industry-sponsored "research" projects at cow colleges.

People then would be forced to make up their own minds and just might decide not to buy food with these things on it. Government and corporate-sponsored efforts to promote these concoctions would create a more level informational playing field. But that's a rabbit trail.

Can't you just see the kindergartner bringing her classmates out to the farm right when Dad is mixing chemicals to spray the corn? "And now, buddies, this guy in the moon suit is my Dad. The reason he's frantically waving at us is because if we breathe the stuff he's getting ready to spew into the air, we could all . . . . ." And she falls over, unconscious. Really friendly place.

The worst is fecal concentration camp animal facilities. Years ago we used to purchase egg cartons from an industry standard, commercial egg farm. We got fairly friendly with the owners and they would let us look around the facilities. We watched them wash all the eggs in yellow fecal water that smelled very similar to vomit. The lady candling eggs told us that the cracked eggs went to a facility that put them all in a huge press--eggs, shells and all. A piston pressed the eggs through a small screen to separate the inside from the shells. This, she explained, was the material used in "egg-less" batters. Our advice is that if you buy batters for cake or any pastry that requires eggs, make sure you have to add your own fresh eggs. Yuck.

The highlight was walking into the laying house. This 50 foot by 450 foot house had three tiers of cages 16 inches by 22 inches, each of which housed 9 birds. The cages were constructed of 1 inch by 2 inch wire mesh. A feed conveyor brought feed along in front of the cages and a little water cup offered drinks. The birds were so crammed in these cages that they could not all sit at one time. Most cages contained one dead bird in some stage of decomposition, gradually being pushed through the screen floor by her living cell-mates. To look down this house at these thousands and thousands of birds, debeaked to keep them from cannibalizing each other, and in a constant state of panic was like looking into something from a science fiction movie. Stephen Spielberg could not conceive of a more morbid, inhospitable place. The stench was unbearable.

We would try to get each of our apprentices to visit there at least once during their year-long stay, and it was a life-changing experience for each one. A couple almost fainted before they could get out. None stayed for more than a few seconds. Honestly, you just stand there in horror and disbelief. This is what modern agribusiness calls "state-of-the-art production." This experience confirmed for us that if the average person saw how their food was raised, they'd become a vegetarian. I identify with vegetarians. If we couldn't get our kind of food, we'd be vegetarians too.

And yet people proudly wearing the mantra "pro-life" patronize these establishments enthusiastically. If they feel a twinge of guilt, they dismiss it because finding alternatives may be difficult, requiring time and monetary investment. But they don't have any problem investing in a golfing vacation or a new growth stock.

This egg factory outfit is operated by two 60-year-old brothers. Not one child is interested in the farm. I wonder why. They are delightful, churchgoing fellows on the outside, running hell on the inside. It's tragic.

Have you ever visited the fecal beef feedlots in the Midwest? I did a speaking tour in Nebraska a few years ago and could tell within about 10 miles every time we were approaching a feedlot. The pall of fecal particulate hanging over these yards and the odor are unbelievable. The manure, both on the ground and hanging on the animals, is beyond description--but I'll try. Imagine a coat of mail hanging on a knight of the Roundtable. Now imagine a coat of mail about three times as thick as normal. Now imagine it made out of manure, and hanging on a beef steer. The kicker is the tip of the tail, which is supposed to have a cute tuft of hair. This tuft of hair provides an attractant for the manure the animals lie in, and often looks about as big as a one gallon milk jug. The poor animal can't even swing its tail with that 10-pound weight dangling from it.

Can't you see the son of the operation's owner bringing his school buddy out to the feedlot? "Hey, Mike, here's our feedyard. See the cows?"

> Mike: "Ahhhhchew! Ahhhhhhchew! Ah cad't see buch."
> Farm boy: "Right there. There are 100 steers in this pen and we've got about 50 pens lined up here."
> Mike: "It stingks. Tade be back to de house."

There's an old joke about a fellow stopping by a feedyard during the spring thaw. He had a cowboy out there checking pens and finally he spotted the cowboy making his way through the cattle. The cowboy was only sticking out of the mud from his waist up. The owner stopped on the access road and yelled across to the cowboy: "Jake, is everything okay?"

Jake replied: "Yeah, boss, but my horse is struggling a little bit."

Anyway, you get the picture. I'll dispense with the pictures, but virtually everything in conventional agriculture is unsightly and stinks. Farmers joke about "country air" when poking fun about the newly-arrived city slickers complaining about the poultry stench. This is one where I'm not sure where to come down. I've always said your freedom ends at my nose. In other words, as long as you mind your own business, or tear up yourself, go ahead. But when you hit my nose, you've gone too far. I'd have an awfully hard time restraining myself if I lived downwind from one of these operations.

I'm not a fan of government involvement, except when your fist hits my nose. And your chicken house stinking up the neighborhood is hitting my nose. The point is, what kind of life is that for your child? To have to defend the noxious fumes and pollution emanating from his farm every day is a terrible burden on a child.

The decision of whether these production models are appropriate, ethical or healthy is really a no brainer. All we have to do is apply the test we apply to everything else living and it will tell us: how does it smell, how does it look? We apply these standards to everything else alive, but as soon as it comes to something like food production, we throw the common sensory diagnostic tools out the window and take on the role of academic egghead apologist. "Uh, let's see. If 20,000 pigs stink up the neighborhood this much, it really wouldn't add appreciably to go on up to 30,000. Nobody could tell the difference. See how efficient we've become?"

Farming models should be sensory-pleasant enough to attract a whole kindergarten class for an afternoon of pleasant memories. Kids should be able to sit on 5-gallon buckets with the chickens and have hours of enjoyable entertainment.

When children visit our farm, we take them out to the Feathernet and let them sit among the chickens. Of course, I often catch a chicken for a child to hold. It's a wonderful experience. Listening to the pleasant clucking of the birds, reaching under a hen on a nest and feeling the warm eggs, petting the soft neck feathers. These are soothing, pleasant, memorable experiences.

Since our children were toddlers, their friends have come out to visit the farm, and I've enjoyed listening to the refrain: "Oh, this is neat. Awesome! This is the bestest place in the whole wide world!" These phrases nestle into a child's heart, generating a feeling of "specialness."

One way to know if your farm passes this aesthetic and aromatic test is what I call the embarrassment factor. When you take a visitor for a walk-around, how many times do you catch yourself apologizing for something? If it's more than twice, you've got problems. No matter how flawless your model and management, something inevitably will go wrong when you're taking guests around. Has to happen.

The cows can be as good as angels for months, but let somebody come to see this wonderful place, and that will be the day a deer runs through the wire or a kink finally gives way and the cows are out. Or it will be the day the water system goes down and the chickens are thirsty. Those kinds of things just happen and they're forgivable. But only two per visit.

Embarrassment about a dead animal, diseased plant, smelly conditions, or whatever, needs to be an exception, not the rule. A corollary to this is the "I'm gonna" routine. If you catch yourself saying "I'm gonna" more than a couple of times, stop and ask yourself why. Don't tell me about "I'm gonna," tell me about what you've done. I'm not downplaying dreams, but anyone can give me a tour and tell me about "I'm gonna." The successful ones will concentrate on what's actually been implemented, what we can actually see going on.

I developed this rule for determining overall farm attractiveness after visiting several farms that exhibited terrible management. The chickens were out of feed and water. The cows were bellowing because they hadn't been moved to a fresh paddock. The porch was covered with duck manure because that was where they roosted at night. The dog (this is one of my pet peeves) kept jumping up on me, putting his dirty paws all over my chest. The nest boxes for the layers had a healthy buildup of manure in them and the eggs were filthy. You get the picture.

The gate sags to the ground, the yard fence has broken boards hanging off it and the tractor sits abandoned on the hill where a tire went flat. The trees in the field where the cows lounge have their bark nibbled off due to lack of adequate minerals in the feed and the barn roof leaks like a sieve. You don't have to possess eagle eyes to notice these things. And when visitors come, the whole tour is punctuated with "Oh, this shouldn't be this way" and "I know this doesn't look good, but I'm gonna . . ." and all the other trite phrases used as excuses to explain away mismanagement.

Minimizing these apologies, and the magnitude of the

unsightly mistakes, is a key component of creating a family friendly farm. I helped a neighbor one time clean up a water spill in his confinement turkey house. A water spill occurs when an automatic waterer fails to turn off when the trough gets full. The water continues to run, flooding an area in the house. This one had been running some time and made a slurry pond about 50 feet across. The turkeys were large, almost ready to go to slaughter.

I went up and used a scoop shovel and 5-gallon buckets to clean up the slurry. The turkeys would get curious about what I was doing and come over to investigate. The slurry was about 4 inches deep and as the big toms strutted over there, they would fluff out their wings and make thumping noises. Their wing tips would drag in the slurry.

By the time I had my two buckets full of slurry, a couple of turkeys had edged really close. One thing that lured them were the rivets on my Wrangler jeans--the shiny copper pocket rivets attracted their attention and they would try to peck them off. Whap, whap, whap! Anyway, I'd hoist those buckets into the honey wagon and the movement would frighten those big turkeys. They would jump and flap their wings, anointing me with beautiful arcs of turkey slurry. Folks, let me break this to you gently. This is not family friendly farming. Kids don't grow up with a hankering to tackle this job. It's not on the dream list. Maybe a nightmare list, but not dream. And that's just the point. Kid's don't tend to be drawn to their nightmares. And a nightmare paradigm, if it lasts one generation, certainly won't last two. Maybe the parents were foolish enough to get into a nightmare, but the kids probably won't be.

In this discussion of specifics, let's not forget the big picture landscape. Pasture weeds, cattle lounging in ponds (we call them hippopotami) and gullies all indicate a production model unbecoming, and unbeckoning to our children's friends. But here at our family friendly farm, visitors are scarcely out of their car when they tug on our children, asking to see the cows, chickens, pigs, garden, ponds, whatever. This makes our kids feel important and special.

Does your farmscape--the land, the buildings, the animals, the plants--soothe you or stress you? When you step out the back door and head to chores, do you feel a calming in your spirit, or an anxiety? I'm not suggesting that once in awhile when we're wrestling with a problem we'll never have some anxiety, but if this continues more than a couple of days, something is way out of whack.

One of the most eye opening experiences for me was a couple of decades ago when I became good friends with a family who had been named "Virginia's Outstanding Young Farm Family of the Year." After getting to know them, I found out that in fact they were head over heels in debt with no end in sight. The father had developed a severe nervous twitch, was running on a couple hours of sleep a night, and the poor mother was trying to pick up whatever slack she could. It was not a pretty situation.

Since being a little boy, he had desperately wanted to farm. He scrimped and saved and finally put together his nest egg. But he built a confinement hog house. Big mortgage. Prices dropped. Disease hit. And as the tears rolled down his cheeks, he said: "You know what really hurts? I did everything exactly like the experts told me to. Everything." It was a moment I'll never forget, a time of deep resolve in my heart that I would never, never go the conventional industrial agriculture way. When I saw the hurt, the devastation, the empty promises, I was furious with righteous indignation over what government schools had done to hardworking, conscientious farmers. Of course, he didn't have to listen to their advice, so it was shared blame. But so unnecessary and painful. He lost the farm. But he followed every USDA specialist's advice . . . perfectly.

We need farms where our children run up to their visiting peers and exclaim: "Come with me. I want to show you something!" You should be able to trust your little ones to take their visitors through any door on the farm and have a safe, mutually enjoyable time.

If we would simply apply the smell and sight tests to our farms, it would fundamentally change the way we produce food in this country. There is no excuse for any other model. Consumers should insist on it by voting with their food dollar. Our farms have become so people hostile that farmers must place biosecurity "no trespassing" signs on the entrances. Visitors can't just drive in. They have to go through a biosecurity check. What does it say about our animals when they are so fragile and immuno-suppressed that we can't even let a class of kindergartners play with the chicks?

When you have to step in a bucket of disinfectant, don disposable boots and disposable jackets just to visit a modern poultry house, what have we done to our food system? Anything this hostile to humans can't be friendly to our intestines. And it sure isn't friendly to our children, or their friends.

Conventional farmers hold to a belief that is axiomatic: efficient production requires mountains of smelly manure, centralized animal factories and a room full of pharmaceuticals. On many operations, we could change the spelling from farm to pharm and it would be more accurate. Sickness should be an anomaly, not a normal occurrence. Sickness is unsightly, unpleasurable, and a real emotional drain. It has no place on a family friendly farm.

The bottom line is this: we can have extremely high levels of productivity without sacrificing pleasant sights and smells. In fact, we can enhance the sights and smells with superior production models. Here are just a few examples:

- Management intensive grazing for herbivores. No feedlots or grain feeding is necessary. The animals are on clean pastures for day long paddock stays.
- Orchard mowing with animals, preferably sheep, along with range turkeys to debug and fertilize.
- Asparagus with turkeys. Turkeys go in as soon as picking is finished, and grow along with the ferns. By Thanksgiving, turkeys have fertilized, debugged, and most importantly degrassed and deweeded the asparagus. The asparagus shades the turkeys during the summer.
- Pigaerators for turning compost. Make a pile, add some whole shelled corn and barley, about 25 pounds per cubic yard of compost, and turn in the pigs.
- Pastured broilers in moveable, floor less shelters. Automatically spread manure and provide fresh air and sunshine. This offers predator protection and high green material intake for the birds.
- Pastured laying hens in electrified netting, sheltered in moveable hoophouses.
- Chickens under indoor rabbit cages to aerate bedding and stir carbon into rabbit droppings.
- Using small grain as a trellis for hairy vetch to increase biomass production in a hay or green manure crop plan.
- Poultry under and between grape vines to mow, debug and fertilize.
- Herb gardens incorporated into rock gardens or water

gardens.
- Edible landscaping rather than ornamental.
- Caged aquaculture utilizing on-farm springs and streams.
- Portable hoophouses and/or greenhouses for winter vegetable production or animal housing.

This list could go for pages and pages, but I hope it shows that by integrating plants and animals symbiotically, we can create models for higher production than industrial paradigms, without any of the negatives. And all the while, these relationships create an aromatic and aesthetically pleasing farmstead.

A family friendly farm must be pleasant to the senses in order to romance the next generation into it. The industrial model cannot do this. But plenty of biologically-based models can and do. The working prototypes are here, on our farm and many others. All we need is the courage to implement them and watch the children shine. The soothing in our own spirits will inspire and call to our youngsters so that they can't ever imagine leaving this paradise.

Pleasant Farms: Aesthetic and Aromatic

Chapter 19

Creating Safe Models

Family friendly farming requires child friendly models. Minimizing the situations that jeopardize safety or sanity is a critical element to make the whole farming lifestyle enjoyable.

No matter how good a farmer you may be, crises will happen. But I suggest that you cannot wake up morning after morning expecting a crisis of the day. That wears thin on the family, making everyone edgy. Things do happen that are completely beyond our control and that should be expected. But these occurrences should be the exception and not the rule. If every day features a new crisis, we're making some major mistakes.

Some farmers just expect bad luck. I know some folks who have another litany of sad tales to lament every time I see them. The equipment breaks down, the animals are sick, the chickens quit laying, the dog runs away and the weeds take over the tomatoes. It's just one sad, hard luck case after another. The amazing thing about this is that seldom do these folks take responsibility for what's happening on the farm. It's just a black cloud that makes nothing turn out all right.

The parents seem to walk around waiting for the next crisis. The children grow up assuming that these tragedies are normal. This results in a pessimistic, jaundiced view of farming, and certainly does not make the children to want to continue such a life.

Crises put everyone in a dither. They overwhelm the family routine, throw off the schedule, and put everyone's sanity on the line. All energy is focused on the situation. Frayed nerves, short temper fuses. Running around helter skelter. Even the best-run farm will have something like this from time to time, but it should happen seldomly. What is amazing to me is the way farmers seem to enjoy putting themselves in this crisis mode.

It's almost like we're a bunch of sadists, trying to figure out how best to put ourselves in a vulnerable situation. Take feeding hay, for example. Here at Polyface, we use the hay shed and feed the cattle through a vertically moveable feeder gate. The hay is all under roof and protected from weather. Furthermore, the cattle come to the hay rather than us bringing the hay to them. Most farmers deliver their hay out to the field to feed the cows.

During the winter, this requires trying to start engines and riding around out in sub-freezing temperatures. "Let's see how cold it can get and still allow me to start this engine," seems to be the mentality. During the spring thaw, when the ground gets mushy, engine starting is not a challenge, but getting around the mud is the new challenge. "Let's see how deep we can bury this tractor and still not get stuck" seems to be the objective. And then when we do get stuck, we can let our crisis envelope the neighborhood with additional tractors to pull ours out of the mud.

In a family friendly system, however, we can feed 100 cows in about 10 minutes without starting any engines at all. The hay shed protects the animals from cold rain and the bedding pack stabilizes their droppings with carbon. New bedding can be added by hand or, when weather conditions moderate, by tractor. We use old hay, straw, sawdust or wood chips, depending on the weather and bedding conditions. It is extremely flexible . . . and family friendly

The anaerobic bedding pack, which stays warm and dry, offers the cattle a clean and warm lounge area, increasing comfort, sanitation and health. The feeder gate protects the fresh hay from the cows' excrement, eliminating parasite and pathogen ingestion and, therefore, the headgating and administering of systemic grubicides and wormers. No matter how good you are at working cattle in corrals and chutes, the likelihood of something going wrong is extremely high.

It may be something as simple as getting your kneecap kicked off to something as critical as getting an animal hung in the headgate, requiring cables, front end loaders and tense moments to free the stressed animal. Plenty of people have been hurt in a pen of calves. It's often by their own inexperience or aggressiveness to be sure, but plenty of times a calf just acts up out of the blue. Why in the world would we feed hay out on the muddy ground where the cows waste half of it by trampling and then spread excrement on the other half that they eat? It's almost like cattle farmers sat down one day to contemplate how difficult they could make their life, and came up with a system that maximized pathogen ingestion, hours in the corral and at the headgate, and unnecessary labor.

A related topic regards calving season. Several years ago I was up in Alberta and Saskatchewan doing a couple of seminars and they were having a once-per-century cold spell. It had been about 80 degrees below zero, Fahrenheit, for a week or so. When I landed in Calgary, it had come up to a balmy minus 30. A veritable heat wave.

They squired me off to the local television station to do an interview for the agriculture program and the reporterette decided since it was so warm we'd go out to the city park. After all, the walk would do us good and it was only about five blocks from the studio. Where I live in Virginia's Shenandoah Valley, we think we know what cold is. Hah! We usually have a week of single digits and I can remember a couple of times we've had a week below zero. But we've got nothing like I encountered up in Alberta.

The cameraman, the reporterette and I headed down the street, me trying to bury my head into the collar of my supposedly heavy winter jacket, and this gal just chattering away about the beautiful day. I thought cameras didn't work well below freezing. This guy must have had a leftover camera from Jacques Cousteau or something. The reporterette told me she had been a Canadian Olympic team figure skating contender just a couple of years before. A knee injury forced her early retirement. At least her explanation revealed why she considered minus 30 to be a perfect day for an outside interview.

We finally got to the park and she decided sitting on a picnic table would be good. Actually, she was sitting; I was shaking. The backdrop was the lake, on which the town zamboni dutifully manicured the ice. She must have been thinking wistfully of her days following that machine around on the solid frozen lakes of the region. I was just trying to keep my brain cells from shutting down.

Somehow we finished the interview before I became

comatose, and my hosts whisked me off to the seminar. I'm used to ten-minute breaks every couple of hours for everyone to get a cup of coffee or go to the bathroom. But every time there was a break, everyone left. I finally asked someone where everyone was going during the breaks. "Out to start their engines," he said matter-of-factly. Oh, dopey me. As if vacating the building to start engines was anything odd.

During the question/answer period, a cattle farmer asked me when calving season should be. I made some joke about it certainly not being during these conditions, and the audience rocked with laughter. I was taken aback. Now it was my turn to ask questions. "Did I miss something? Why was that so funny?"

"Because we're all calving now," replied a pensive cattleman.

"No! Tell me it's not so," I said.

"Oh yes. Most of us have at least a spouse or child at home today keeping heat lamps on calves in birthing stalls, trying to keep them alive. If we don't get them cleaned off within 60 seconds, they freeze to death. And we can't shut off our tractors lest we be unable to get hay delivered to them tomorrow. So we just leave our tractors running 24 hours a day."

I was dumbfounded. This was mid-January. Who in the world would be calving this time of year? In Alberta, no less! Of course, I'm familiar with so-called "spring" calving in Virginia in January, and I know how silly that is. But this wasn't Virginia and these weren't southern boys. The whole interchange pointed out just how much farmers seem to collectively decide to do things that complicate their lives and cause them to lose money.

It's almost like farmers sit around in discussion groups, excitedly inventing new ways to make their lives miserable. "Isn't there some harder way to do this?" is the icebreaker question. From there, everything flows downhill.

Let's take another example of potentially hazardous situations--land clearing. Converting overgrown, forested areas into pasture normally requires heavy machinery and lots of mechanical mowing. We use pigs. Some folks use goats. In either case, it eliminates children watching Dad's reaction when a sapling staub runs through a tractor tire. These are the experiences that need to be minimized. Compare that with watching pigs or goats do the work, and you can appreciate how much less stressful the animal working models are compared to the machinery models.

In fact, machinery itself is a major factor in high-stress situations. The funny thing about breakdowns is that they always happen when the machine is being used. Breakdowns don't happen when the machine is parked quietly in the equipment shed. If we could just keep from using it, we wouldn't have so many breakdowns . . . and unsafe situations

But giving ourselves some wiggle room when we do use equipment keeps us from pushing it to its limit all the time. On a family friendly farm, for example, by grazing all our fields with the cattle early in the spring, we delay physiological maturation of the hay and shift our haymaking season toward early summer. We can make good quality hay nearly a month after our neighbors because we've grazed it once or twice before letting it go for hay. This allows us to make May quality hay in June, when the weather is hotter and drier.

Rather than fighting the early spring cool temperatures and

frequent rain showers, we make our hay when conditions are more conducive to getting it up in good shape. This management technique takes some pressure off while we're working.

Keeping plenty of hay in the barn gives us a safety valve, a ballast, against drought years. I can't believe how many farmers zero out every year on their stock and stored forage supplies. One rough drought, and these guys are screaming for government assistance to help them through. Instead, we like to keep at least half a year to a year's worth of hay stored under roof just in case we have a bad year. If this buffer slowly increases, we know we can add a few more cows. If it slowly diminishes, we know we're a little overstocked and need to drop our numbers.

This forestalls panic when everyone else is going crazy. Don't get me wrong. We don't like droughts any more than anyone else, and they always hurt. But if you plan for them and don't run everything right up to your eyeballs, you can take the valleys calmly and confidently.

This is why drip irrigation in a vegetable operation is imperative. You must realize how emotionally devastating it is to you and the children when they see a crop wipe out. All that time and energy, and it's just gone. The point is that droughts, floods, disease and pestilence will happen. Period. It's the nature of nature. Part of our jobs as farmscape architects and resource stewards is to build forgiveness into the system. We must actually plan and design into our scheme protective models to buffer nature's severity. To go along innocently and assume that this won't be a rough year is foolish. It's also setting us up for a big disappointment.

And big disappointments aren't lost on the kids. They know when we're in crisis. Try as we might, we cannot shield them from the emotional and economic pain inflicted by these calamities. On a family friendly farm, you should be calving when the deer are fawning. Mowing hay when the weather is better. Installing driplines in the vegetable operation. Feeding excellent minerals and maintaining sanitary systems reduce cattle working requirements. Building ponds to store more water runoff.

Every spring we used to run the yearling calves through the headgate and squeeze out heel fly warbles by hand. We kept a record of which calves had the most and culled those mama cows. We made steady improvement over a number of years. But when we began using the eggmobile as a cattle hygiene system, we suddenly had no more warbles. None. Nada. And we quit having to run those calves through the headgate. One more detestable ritual extinguished. It's safer for us and less stressful for the cattle. Compare running the cattle through the headgate to gathering eggs. Which one would your children rather do?

Our entire American farming structure has become child unfriendly. Consider the grain crops in the Midwest. Seventy percent of this grain goes through herbivores, who are supposed to eat forages rather than grains. Not only do we take the vegetative perennial prairie skin off the land, opening it up to erosion, but we also eliminate the livestock, which we coop up in fecal feedlots. Trucked to centralized processing facilities which are unfit for humans, let alone children, the animals eventually wind up in supermarkets around the world. How involved can children be in this whole process?

Compare the current paradigm to a system in which the animals grazed on the land under management intensive grazing. Teresa and I had to go receive a conservation award when Daniel was about eight years old. At the time, we had right at a hundred head of cattle. Before Teresa and I left, I laid out the next three paddocks with electric fence. Each day, while we were gone, he moved the herd into the next paddock. It is quite impressive to trust a small child to move a hundred head of cattle by opening a gate and calling them. Talk about leverage. To see all those thousands of pounds begin to move as one unit at the behest of a little child is a powerful experience.

Instead, America has a grain-based herbivore production model full of hazardous machinery, dangerous grain buggies and augers. Certainly children can't be involved with every facet of every production or processing system, but the safer and more child friendly we can design each part, the better.

Just a word about marketing. Farm children typically have no connection between what they produce and the end user. How much more meaningful it is to actually be a part of that consumer appreciation process. Our marketing efforts, rather than eliminating the part children can play, should encourage active participation and integration by children.

Overall, a family friendly farm should minimize risk from the vagaries of weather, disease, price and pestilence by utilizing diversity, irrigation, feed storage, greenhouses and other strategies. The production, processing and marketing models should incorporate rather than exclude children by being child friendly and child safe. And we should farm in such a way that we reduce the chances of crisis situations, from wild calves to machinery breakdowns. The question: "Is this something I can do with my kids?" should permeate every decision we make. If food is not child friendly, it's probably not nutritionally friendly. The whole is only as good as the sum of the parts.

Creating Safe Models

Plenty of techniques exist to accomplish child friendliness in farming, including minimizing parental anguish. Suffice it to say that the conventional model does not take our little people into account, let alone big people. But a family friendly farm sees the child friendly element as crucial to an enjoyable and happy environment.

Chapter 20

Multiple Use Infrastructure

Diversity is quite a buzzword in our society right now. Environmental diversity, cultural diversity, diversified investment portfolios and ideological diversity. On the family friendly farm, we must also consider infrastructure diversity.

By that I mean the buildings and machinery, as well as the landscape. Diversifying the landscape goes without saying. Forest, open land, riparian areas all contribute their plant and animal species to the ecosystem. Diversifying the landscape makes a more interesting and picturesque place to play and work.

Think about how a pond mesmerizes children. It grows tadpoles, salamanders and frogs. Big bullfrogs wait until you're inches away, then croak and jump, scaring the living daylights out of you. If you haven't paddled a canoe with your kids, you're missing out on some good times. In the winter our ponds offer ice skating.

Creeks. Oh, what fun creeks can be. Building dams, flipping over rocks and searching for crawdads. Splashing and swimming in a mountain creek is about as good as it gets.

And then there is the forest, with its deep, dark, foreboding innards. Those massive trees that groan and creak in the breeze. The chattering squirrels dropping nuts that frighten, not to mention the occasional grouse that flies up one foot in front of you. As kids, we built many a fort against the upended base of a tree.

The open land, with its waving grasses prior to haymaking, opens the landscape, providing panoramic views. Early spring grass is the best playground surface, and beckons all children with that ephemeral experience of rolling around in soft green sod. Certainly a diversified landscape is a curiosity-inducing, interesting place to live.

But beyond that we must design multiple use into the improvements. And here is a critical element in the family friendly farm. The law of physics that for each and every action there is an equal and opposite reaction is true emotionally as well as physically. Massive, single-use, capital-intensive buildings and machinery shackle the next generation. And what does a shackled person want to do? Flee.

Nothing gives our children a greater feeling of bondage than confinement livestock facilities. No doubt, on this topic, this is the worst offender. Closely related are things like combines, capital intensive waste management systems and big machinery. But let's stay with the confinement animal facilities for now since they are the prime example. At one time, the county where we live had a dozen or more confinement hog facilities. Today, not a single one of them is operating, and to my knowledge, those pieces of real estate are no longer owned by those farmers.

What happens in the life of a confinement facility, and it really doesn't make any difference whether it's dairy, beef, broiler, egg, turkey or rabbit, is that the farmer takes out a mortgage on the building. For sake of discussion, let's say it's a 20 year mortgage for $200,000. The life expectancy of the house is probably about 20 years. Remember, these houses are usually saturated with acidic ammonia year-round, which eats away at metal.

Animals go into the facility and the grind begins. Hauling manure, hauling feed, veterinary care and building maintenance. The children grow up with this routine, thinking this is what farming is--haul, haul, haul, fix, medicate. And all the while, the bank wants its money. Of course, this system is far more tenuous in a vertically integrated industrial setup such as poultry than in an owner-owned, animal-owned operation like a dairy. Under vertical integration, the farmer doesn't even own the animals. He just owns the land and building and the integrator supplies animals and feed.

Now let's assume the children hit early teen years and want to do something a little different. Maybe one of them wants to have a chicken project on the side and sell eggs to the neighbors. Poultry contracts preclude their growers from having any chickens on the farm except those supplied by the integrator. The child's little side business is out of the question.

But beyond that, what if the child wants to make a change in the system? Or maybe a new model offers tremendous opportunities. For example, seasonal, grass-based dairying rather than confinement. The problem is that the older generation has invested so much time, energy and money into the existing facility, that it is too difficult to abandon. And the payments keep the farm strapped for cash so that even when the facility becomes obsolete, upgrading isn't possible.

In the case of the vertical integrator, whenever a new watering or feeding system comes along, the company requires the farmer to retrofit the house (at his expense) or lose his contract. Of course he's going to install the new system because he can't afford to lose the production contract since he has to keep making those payments. Even if he gets tired of raising chickens, he can't get out of it because the bank wants their money and the only way to generate the dollars is to keep that house full of what it was designed to produce.

And here is where the bondage comes in. The farmer feels stuck. The indebtedness reduces the economic flexibility on the farm. The economic straightjacket creates an emotional straightjacket. And that's when everyone begins to hate what they're doing.

Many poultry farmers build their houses under the promise of netting $15,000 a year only to find that it's $10,000 or less. They hope they can beat it by building another house. But every other grower is thinking the same thing, so because the contracts rank growers against each other for compensation rates, the true figure drops to $5,000 per year. At that point, the farmer says: "Forget this. With $400,000 I could invest it in Treasury Bills and make more."

But he can't get out. The bank wants its mortgage. He can't sell the houses for the indebtedness. He doesn't want to sell the land. And so he continues putting chickens in, batch after batch, a serf to the feudal lord. And he did it to himself, completely voluntarily.

One of my favorite farmers is Paul Bickford in Wisconsin. He is one of the country's most successful dairymen. Actually, he would say his wife is the most successful. What she's done with a little grass-based Jersey dairy is nothing short of phenomenal. But that's another story. The point I want to make is that Paul and Sidra Bickford had all the degrees and financial backing to put in the latest and greatest.

They poured concrete and built blue silos (I call them "bankruptcy tubes") and were successful enough to be featured on the front cover of every leading dairy journal in the country. He was the darling of the industry. He had 12 employees, milked hundreds of cows, and was so successful he was able to amass a couple million dollars of debt. One day he just decided to do something different. He didn't know what, but he knew there had to be a better way. Their cull cow rate jeopardized their replacement sustainability. He went to a conference on grazing and came back ready to begin grass-based dairying.

It wasn't easy. In fact, today he's very up-front that if he had it to do all over, he would have sold the entire farm, lock, stock and barrel, and started over somewhere else. But he didn't. He slowly began making the changes. And after almost a decade he's almost debt free, loving life and farming. He and Sidra are a poster family for successful turnarounds. But his statement that is most powerful to me is this one: "I can't afford to fill those silos. Every time I begin filling them, I'm losing money."

And there they sit, "monuments to the stupidity of man," to quote one of my favorite phrases from Charles Walters, founder of ACRES USA magazine. The problem is that the capital-intensive, single-use machines and buildings generally cannot be economically retrofitted to other uses. Even if they outlive their usefulness; even if that whole model becomes obsolete, the farmer feels emotionally and sometimes is economically compelled to keep using them. This does not encourage family friendly farming.

Some people think I'm real creative, and sometimes I like to think so. But let me relate a story to help illustrate that I have gone through this same mental shackling. Years ago when we began the apprenticeship program, we began large-scale pastured egg production. These two projects developed at the same time. We built portable shelters just like the broiler shelters we'd used for years, but tucked nest boxes under one end. That tripled the amount of time it took to build a pen because the nest box installation was tedious. We spent hours and hours building all these shelters--29 I think we had.

Each pen cost about $200 in materials alone. They were pretty to look at out there in the field. The total infrastructure cost was $6,000, not counting time, which was at least that much again. Oh, we took pictures of it, and I added it to my slide program for conferences. I just thought this was the cat's meow, the slickest thing since sliced bread. The apprentices moved the pens each morning and we went along with this model for a couple of years.

Michael Plane and Joyce Wilke visited from Australia, a couple who had seen my writings on pastured poultry and wanted to stop by our farm since they were in the States. They are fans, as I am, of Eliot Coleman and Bill Mollison's permaculture concepts. Coleman is famous for his multiple use greenhouses that roll on tracks. Quite ingenious.

Anyway, Michael told me I should dispense with the small field shelters and use a brand new electrified poultry netting being made in Britain. He said he was using it and getting along fine. He said it would reduce our labor, give the chickens (chooks) more grazing and thereby reduce feed costs and produce a superior egg.

I listened politely during the delightful visit but I just couldn't imagine anything better than what we were doing.

After all, Australia didn't have foxes and coyotes. And what in the world would we do with the pens? We went on for two more years until Andy Lee, of Chicken Tractor fame, and I were talking one day and I asked him if he had any experience with this electrified poultry netting. He said he tried it and it wouldn't work. That was just the impetus I needed. A little friendly competition never hurt anyone, you know.

I called Premier Fencing and ordered a roll. We put it around one of the field shelters and turned the chickens loose in this electrified netting circle. To our complete surprise, they stayed in. And predators stayed out. This went on for two weeks. I started getting excited. We quickly fabricated an A-frame shelter, ordered some more netting, and put a couple hundred birds out. It was fantastic. It was everything Michael and Joyce said it was.

I told Andy and he pulled his old piece of fencing out, tried it, and got along great. We had a good laugh about it and he then continued experimenting. We invested in more fencing, built a couple of hoop shelters, and pondered what to do with all those individual pens we had used before. I drug my heels for two years. This paradigm shift would completely revolutionize our pastured egg enterprise. But I wouldn't do it for two years because of the economic and emotional investment in the previous paradigm.

Now I ask, if I felt that reluctant to change something as simple as this, how must a person feel who's looking at a quarter-million dollar confinement livestock facility? The emotional attachment is real, aside from the economic. When we spend that much time in a structure, creating that many memories, we just become attached to it. The idea that what we've invested our life energy in is obsolete threatens and depresses us. No one wants to think their model farming system is outdated. And no one wants to think their life has been inefficient and unproductive.

Building diversity and flexibility into the farm infrastructure insures a freedom not only for the parents but especially for the children. That is one reason I like pole buildings. They can be built cheaply and utilized in numerous ways. For example, our hay shed for winter feeding performs the following functions:

- Hay storage
- Carbon storage for bedding--sawdust, wood chips, straw
- Cattle housing in winter
- Manure and urine protection
- Compost protection
- Pig housing
- Rabbit housing
- Chicken housing
- Sheep housing
- Machinery storage
- Picnic shelter during field days
- Play area for children

All of this happens at different times of the year, but look at the multi-functions of the one structure. Not only is it being used year-round, but the multi-use breaks up pathogen cycles and makes the whole structure more interesting.

Dad was big on portability, partly to preserve flexibility. Always ready to try something new, he wanted to leave himself an out. Easy exiting is one of the big secrets to freedom. When we can't see the exits, we feel panicky, hemmed in. But as soon as we see the exit sign, we can relax. Not that we wanted to get out necessarily, but because now we know we can get out if we need to. This is the feeling I'm talking about preserving for the next generation.

One of the most liberating things we can do for the next generation is to offer an easy exit. Nothing is more shackling to the next generation than to look at their future and think they have to spend it doing exactly what Mom and Dad did. What young person wants to feel that way? Even if what Mom and Dad are doing is fun, most young people have aspirations to put their own stamp on the business, and we need to make sure they see an open door to accomplish that.

One of the reasons we have our apprenticeship program is to preserve the open door for our children. The prototypes we've perfected require little daily creative energy now. Our mature prototypes have a protocol that works, but their productivity still requires thoughtful, caring warm bodies. In order to make room for our children to add their own creative prototypes, the apprentices maintain existing prototypes. The apprentices don't have the benefits of our children, who grew up with all the trials and errors. Our kids understand the whys and wherefores of each model, and can articulate why it looks the way it does. The uninitiated ask some of the funniest questions or have some of the most harebrained ideas, but it's because they haven't been through the fire.

This meets the needs of the apprentices for hands-on experience with the models, but does not entrust the research and development to them. As their competency increases, they often add their own refinements or additional models to the farm mix. If we started the apprentices at the prototype development stage, they would not have the advantage of knowing how and why the successful models work. Learning and discovery move incrementally from elementary to complex.

Of course, sometimes a person coming from outside has a wonderfully creative suggestion, too, because he has a whole fresh perspective. The point is that if we maintained the same level of production for everything, the children would be simply working in the mature prototypes developed during their childhood. The apprentices, by maintaining those mature and therefore lucrative prototypes, not only learn their intricacies, but also allow our children the freedom to explore additional prototypes.

I'm not suggesting that apprentices are necessary on every family friendly farm. But on this one, where I now spend half my time in information dissemination and education, we've lost half a farmer. I could quit writing, lecturing and conducting on-farm seminars and return to full-time farming, but that doesn't fit our goal of letting each person function to their fullest potential. And so we wrestle with maintaining this balance, but the wrestling is the stuff of life. Perhaps a day will come when apprentices will be obsolete here. That very well may be, but for now, they play a specific and important role in this multi-generational freedom feeling.

As another example of infrastructure diversity, our brooder houses are on skids so they can be moved around. Always a person for function and always finishing a brooder house a day before we needed it, I just parked them willy nilly. When Rachel turned 13, she decided it was time to make the place look like something more than a dump. As Teresa says, "can we at least make it look like somebody lives here?" Ouch.

So we all sat down, sketched out the farmstead buildings, and let Rachel begin rearranging. Amazing. In an hour, we had a far more functional, aesthetically pleasing farmstead plan. So we moved all the brooders around, and sure enough, everything looks much more attractive, and it's far more functional. When she has that much say over things, and then watches all the scurrying around to make it happen, that's a wonderfully empowering, liberating feeling. This creates in the kids the knowledge that they can change things.

One of the fencing rules in controlled grazing is not to install any permanent fencing at first. Whatever you don't move for three years, convert to permanent. That is not a bad rule, and it speaks to flexibility. Be slow to buy or build things that become permanent fixtures in the business or on the landscape. For example, we ran out of equipment shed space and parked the extra machinery in another spot across the lane. Essentially we parked things in a line on this spot and it looked kind of unkempt, like trucks and machinery look when they're outside.

Finally, we built a shed on that spot. Now it looks nice, but basically we just roofed over the area where we always parked. It wasn't a change of use. The land use pattern became the defining factor in what should go there. As a result the new shed blends in perfectly and is highly functional. I've watched people buy a piece of land and immediately begin building things here and there and everywhere. A friend bought a farm and immediately installed $20,000 worth of permanent woven wire fencing. Then he learned about controlled grazing and asked me to come up and help him with field design. Of course, he had all these wonderful square fields that adulterated the landscape terribly. But the fence was in and it was too late to change. A real shame.

This happens more often than you might think, partly because our desire to build exceeds our experience or our observations. Instead of building roads all over the place, why not make do for a couple of years. Wherever you've worn the grass down, install a permanent road. Then you'll let your experience, and the natural travel patterns, define where things should go. This contributes materially to a family friendly farm.

We do a fair amount of logging and people are always coming by telling us we need a skidder and knuckleboom loader. I admit we're slower because we work with forks on a tractor front end loader, but that is a multiple use machine. We can pull a trailer, hay baler, take the forks off and use the loader for rocks or sawdust. And how do we transport the log? In a dump truck. Yes, it's a little slow to load the logs up over the sides, but again, the truck can haul firewood, compost, wood chips, dirt, gravel or whatever.

What can a combine do besides sit most of the year and then harvest grain for a couple of months? You should always be dubious about machines that sit 90 percent of the time. For all their efficiency at harvesting grain, I'm sure many farmers would do better to have a tractor-pulled machine. Anyway, I think you get the point.

Knowledge rich farming is also a way to reduce single-use, capital intensive infrastructure. Letting the animals do the work, grass-based production, and direct marketing are all knowledge dense. Knowledge does not have to be borrowed from a bank. Knowledge does not increase indebtedness, unless it's knowledge in conventional thinking. The animal and plant integration we practice on our family friendly farm allows us to accomplish far more work, including hygiene and sanitation, on much less capital.

Just consider the eggmobiles, which sanitize the grazing paddocks behind the cows. Or the vineyard with turkeys underneath. Or the pigs that turn the compost. Or the chickens underneath the rabbits. All these functions normally require single-use, expensive machinery. In our case, economically appreciating livestock does all the physical work. The hardest work is the time we spend thinking about how better to integrate the different components, how to close the loops.

Family friendly farming creates a field of new opportunity for the next generation, rather than a field already fully developed. Offering exit doors, multiple use machinery and buildings, and time to be creative are the essential requirements to maintain the freedom. When our kids feel this freedom, their natural desire to explore, to discover and to create kicks in. This in turn fuels the multi-generational train, empowering it to roll confidently and aggressively down the tracks of life.

Multiple Use Infrastructure

Chapter 21

Complementary Enterprises

ACT I

The stage opens on a farm family--Mom, Dad, Junior and Juniorette eating dinner.

JUNIOR (17) Dad, I've been thinking. Do you think there's enough money here to make room for me?

DAD (43) I sure wish it could.

MOM (42) *interrupting* I've been going over the books, and we're just not getting ahead. Once I get the taxes paid and the payments on that new silage wagon, we're scarcely going to have enough left to replace that worn out old Allis Chalmers.

JUNIORETTE (15) I hope in the next couple of years the farm can support at least one of us. What will you guys do without us around to help?

DAD (*laughing*) Boy, you kids talk like Mom and I have one foot in the grave and the other one on a banana peel. Look, there's nothing we'd like more than to have you kids here on the farm, to have you raise our grandchildren right here. But you must understand something--

MOM Be gentle, now, dear.

DAD This farm is doing better than most. Neither your

mother nor I have had to take a job in town to stay here, and that's unusual for a farm this size. But with prices the way they are, and these environmental regulations, higher cost of fuel, machinery, higher taxes--it just goes on and on. One day you'll understand. It's just not as simple as it seems.

JUNIOR (*eagerly*) Look, Dad, I've been reading this magazine about alternative stuff. Maybe we could do something different. Have you ever thought of growing Christmas trees? We've got that eroded hillside back there behind the knoll. There ought to be something we can do.

MOM (*sadly*) Junior, we've been over this so many times. We need every acre we can get to feed the cows. We can't afford to just take acres out of production. We're barely able to make it as it is.

DAD Look, the safest thing to do is to apply for some college scholarships so you can get a good job. I have no idea how this farm could possibly support another salary. It's barely doing one. I just don't know if the future is here. But if you guys saved money, maybe you could buy that old Nufer place and then we could double the cow numbers. That would sure make a difference.

JUNIORETTE But I love the farm. I don't want to go somewhere else. I hate the city. Expressways, house lots . . . But I guess if that's the only alternative, that's what I'll do. (*dreamily*) Maybe I'll marry a guy who's on a big spread.

MOM (*laughing*) That's what I did.

DAD Yeah, right. Listen, we really appreciate that both of you would like to come back. But this farm just isn't big enough to support two salaries, much less three. We don't know what the future will bring, but one or both of you could land really good jobs. If you do well financially, maybe you can put togther enough equity to leverage it onto the Nufer place. He's pushing 60 and none of his kids are even interested in the farm. In another 15 years he just may be ready to make a change. When we retire, we'll be ready.

JUNIOR (*exclaiming*) But that will be 20 years!
(CURTAIN--END OF ACT I)

ACT II

(Curtain rises on Thanksgiving Dinner. Mom, Dad, Junior, Juniorette, their spouses and two preteen children apiece)

DAD *(obviously aged) (clearing his throat)* Junior, Juniorette, do you remember that conversation we had . . . let's see, maybe 20 years ago, about you're coming back to the farm?

JUNIOR Yeah, I think I remember. At the time, I thought maybe we could all work together. But that's been a long time.

MOM Remember, we talked about the Nufer place maybe coming up for sale sometime. Nufer didn't wait very long, did he? He hit 67 and sold it to that city slicker from New Jersey so fast nobody even knew it was on the market. Got a fancy price, I hear. That city slicker. . . . what's his name, dear?

DAD Oh, you mean Walt, Walt Henry.

MOM Yes, that's his name. It's hard to keep up with all the changes. And we don't get around like we used to.

DAD Well, I've been thinking. I'm due to start Social Insecurity next month, and we'll have two eligibilities by the end of next year. We've managed to save a little bit and that will help. I just can't keep things up like I'd like and it won't be long before I simply can't do it. Would one of you be interested in coming back? We'll just do on less but I think it would be a great place to let the kids run.

JUNIORETTE'S DAUGHTER Eeewww, what would we do all day?

JUNIORETTE *(scolding)* Now, Honey, that's not nice. Why, a farm has all sorts of things to do.

JUNIOR'S SON *(laughing sarcastically)* Sure, like bale hay, chop thistles, step in cowpies---

JUNIOR *(embarrassed)* That's enough. I loved the farm

when I was growing up. Look, Dad and Mom, here's the situation. I've been with the company now 15 years. I've had several promotions, and the next one will be really big. We've almost got our house paid for--

JUNIOR'S WIFE And then we can go to Hawaii! Oh, we've been looking forward to that for so long. Besides, I wouldn't know where to begin on a farm. I'd probably ruin the whole thing.

JUNIOR'S DAUGHTER And how would we ever go to McDonald's or Taco Bell? We'd never get to hang out at the mall. That'd be the pits.

JUNIORETTE Well, we made a commitment that when the kids were small, I'd be a stay-at-home mom. It's been tough, but we've managed. I certainly liked the farm then, but now the kids are in soccer, ballet, and little leagues in the summer. It would be terrible to interrupt that. And I sure didn't marry a farmboy.

JUNIORETTE'S HUSBAND If she stays at home, that would put me as the farmer. Oh, boy, that would be a joke. I don't know a bull from a ram. Remember how I jerked the tractor last time I drove it? No, the farm's not for our family.

DAD Realistically, when do you think either of you would consider living here? Or would you at all? The fact is, we just can't hold it together forever.

JUNIOR I've always had in the back of my mind that I'd retire here. Get out of the rat race. The kids would be gone. I could live on retirement so the farm wouldn't have to generate a lot of income. It would be a really restful place.

DAD When would that be?

JUNIOR Well, let's see. *(figuring)* It would be when I'm 63, and I'm 37 now.

DAD AND MOM (gasping) That's 26 years!

(They simultaneously lower their chins and stare at what's left of the Thanksgiving meal.)

(CURTAIN -- END OF ACT II)

ACT III

(Curtain opens with Junior and Juniorette, obviosly older, carrying furniture out of the dining room)

JUNIOR Sure is a shame, isn't it?

JUNIORETTE Remember when we used to play checkers on this table? Remember all the discussions we had?

JUNIOR Yeah, especially the ones about us coming back to the farm. That was really hopeless, wasn't it?

JUNIORETTE I don't know why we ever let ourselves dream such things. Because we were kids, I guess. I wouldn't begin to come back here now. Goodness, we're talking about selling our big house and trading down now that the kids are gone. It'll be nice to have at least a couple of these things from the farmhouse, though.

JUNIOR No way I'd come back. I'm burned out at the company, but it sure gives me a nice paycheck. It'll be good being rid of this place. I'm glad Dad and Mom went so close together and didn't eat up all the estate in the nursing home. Took all their savings, but that was about it.

JUNIORETTE I wonder how long it will take the farm to sell?

JUNIOR Don't know. Jack down at the realty office thought it would go pretty fast. He says a lot of baby-boomers are looking for places like this.

JUNIORETTE The auction is next week. It should bring enough to pay for the last of the medical and nursing home bills, maybe the funeral expenses. Do you think there will be enough left over for us to get anything?

JUNIOR I doubt it. That nursing home was expensive, but I wanted them to have the best. They sure went downhill fast in there. You could see them waste away from visit to visit.

JUNIORETTE This farm was their life. Dad never had a hobby. Mom didn't know what to do without her canning. Well, they're better off now.

JUNIOR Yeah. I'll be glad to be rid of the whole mess. When the real estate sells, let's splurge on a first class trip to Hawaii for both of our families. The kids will all be married soon and it might be the last opportunity to have a good fling, all of us.

JUNIORETTE That sounds fun. I'm sure we'll put whatever's left in the stock market. Sure beats trying to farm. Remember how Dad used to complain about farm prices all the time? Never thought it would be good. *(laughing)* Just sometimes less bad than others.

JUNIOR I think we'll finish paying off the kids' college loans. Might buy a pick-up truck. I've always wanted one of those.

(Curtain closes, end of ACT III)

That play, dear folks, is recorded daily in the farmsteads across America, found under the heading: **REAL LIFE TRAGEDY**. This is one of the most common dramas occuring on the farm scene. It has numerous variations, but the basic plot happens on thousands of stages per year.

The basic problem is that at the time when the young people are gung-ho and ready to come onto the farm, there's no salary for them. What we desperately need in agriculture today are non-competitive, complementary enterprises that allow the next generation, whether related to the elderly owners or not, to create their own salaried enterprise without pulling income from the existing enterprise. The good news is that these enterprises exist in more ways than you can imagine.

The critical period is the time between the kids hitting late teens and when Dad and Mom retire or otherwise do not need farm income. That period of time averages about 20 years, which happens to be the extremely high energy period of the second (shall we say, replacement?) generation. The way it's supposed to work is that the younger generation comes on just as the older generation acquires significant experience. Then the older generation leverages that experience with young energy to create a synergistic business combination. Like I said, that's the way it's supposed to work.

The problem is that farmers generally never think about complementary enterprises. Dairy farmers can only think in terms of milking more cows. That takes a greater land base, more machinery, more buildings. A beef cattle farmer thinks in terms of more cows, creating the same problem as the dairy farmer. A crop farmer thinks about more acreage under cultivation.

In all these situations, the capital is controlled by the older generation, but is only being utilized a small percentage of the time. A guy can only drive a tractor so many hours a year. The goal on a family friendly farm is to let the new generation utilize the existing capital (land, buildings, machinery) to greater capacity in order to keep start-up costs low. But the enterprise must not compete for existing capital. Adding beef cows to a dairy operation is adding cows to cows. That's competing.

Adding cherry trees to apple trees is trees to trees, a competing enterprise. Cherry to apple, beef to dairy is diversity, but it's not complementary diversity. If Dad is growing apples and Junior wants to grow cherries, the two enterprises will compete for the same land base. Not a good situation. Most farmers have never thought about complementary enterprises, so let's explore a few of them.

A few years ago I did some farm consulting. The money was great, but the emotional frustration was too high. Anybody who could afford me wouldn't do anything I said because it wasn't country club appropriate, and anybody who couldn't afford me didn't need to spend the money. They just needed to buy the books and start with their own adaptations. It's all in the books, believe me.

Anyway, I would routinely go consult for beef cattle operations. The first item of business was to get a handle on total animal units per year and do a manure audit. That's right . . . a manure audit. A cow drops about 50 pounds of goodies out her back end every day. These farms normally had some silage, maybe were buying in some protein or crimped corn for energy, and fed their own hay. Otherwise, it was pasture. Commonly, our sleuthing showed that these farms were producing five tons of cow goodies per year. Then I'd flip over to the fertilizer budget and they'd be spending $15,000 a year on fertilizer.

We're not counting the chemicals, like herbicides and pesticides. Just fertilizer. The same 50-year-old farmer would usually have a son or daughter wanting to farm, but all they heard was the proverbial "this farm can't pay two salaries." Clearly, the cow-generated nutrients were not being utilized. Generally, they vaporized off into the air or leached into the groundwater because of inefficient management. These farms could have easily put a child in charge of a composting program and other nutrient cycling techniques, eliminated the fertilizer bill, and thereby created a full-time salaried position.

Sound radical? Maybe, but this is the kind of thinking necessary to make existing farms multi-generational. If we wait to bring on the next generation until after Mom and Dad are off the scene, the business and all its key elements--production, experience, cash flow--loses continuity.

Assuming the kids are on your team, this whole issue is probably the single biggest make or break point in the multi-generational business. It's probably the most common weak link in the continuity chain. This is probably the singlest greatest contribution our farming prototypes here at Polyface have made to the needs of the greater farming community.

Let me describe one of our fields and its production.

We move pastured broilers in floorless shelters across the field at a production density of about 500 per acre. Moved every day, these 50 shelters cover one acre per week. This generates about $3,000 per acre in gross income, $1,500 net for a total on 20 acres of $30,000 annual net during the six month growing season. Six months on, six months off. The broilers fertilize the pasture, producing lush, highly productive forage for the beef cows.

Because the broilers like short grass, we mow it ahead of the shelters with cows, or what we call salad bar beef. The cattle require the least labor per acre, but also return the least, at about $300 per acre gross and $200 net. The cows drop their manure, which incubates parasites and pathogens. The cattle mowing also uncovers grasshoppers and crickets.

We follow the cows with the eggmobiles, a two-house train of portable laying hen houses. Nearly 1,000 hens free range out from the eggmobiles, scratching through the cowpies to eat fly larvae and incorporate the fertilizer into the soil surfáce. Along with the sanitation process, they harvest the newly uncovered insects. The eggmobiles gross roughly $500 per acre in annual egg production, with a net of $300. I'm not adding anything in for the fact that this procedure eliminates toxic parasiticides and headgate working time on the cattle. These hard-to-quantify benefits are just cream.

In the fall, we grow Thanksgiving turkeys in these broiler shelters, yielding another fertilizer application from yet a different species. Each species concentrates a different ratio of nutrients in its manure. The various species balance out each other in the type of forage they like to consume, the nutrients they apply to the soil, and pathogen host cycles. The turkeys gross about $1,500 per acre, a net of about $800.

The cows go across the field up to six times during the year, depending on rains and hay schedules. The eggmobiles follow the cows in each rotation. The broilers go across once and the turkeys go across once. If we add up all the net incomes, we see $2, 800 per acre. The same equipment serviced all the enterprises, same buildings and same land. I realize we were importing grain for the poultry, but the world has too much commodity grain anyway. That cost is all figured into the net.

My point is that on 20 acres we've netted $56,000. And this whole scenario is essentially only an 8 month deal, not a full 12. We do this year in and year out, so I know it's a proven model.

Each enterprise on a family friendly farm returns enough to produce a salary in its own rite. But beyond that, the byproducts of one helps the next one. The broilers and turkeys add manure which grows more grass. The herbivores harvesting the grass create a succulent short-grass table for the following chickens or turkeys. Closing the nutrient and pathogen loops synergizes the production per acre, creating enough net income to support additional salaries.

Now let's move to the hoophouses, where the pastured laying hens go in the winter. The hens live comfortably on the floor and

Daniel's rabbits are in cages up above. This stacking is a linchpin concept in Permaculture , developed by Bill Mollison. The chickens scratch through the rabbit droppings, aerating the area so that it composts and doesn't fumigate the rabbits with ammonia. The rabbits are healthy as a result.

The chickens lay eggs in their spacious nest boxes and fluff in the carbonaceous bedding on the floor. In the spring, the layers go back out on pasture and we can grow vegetables in the composted bedding of the hoophouse, which has been debugged by the overwintered chickens.

In addition to the eggmobiles, we have two feathernets, which is our term for 1,000-hen production units surrounded by special electrified poultry netting. This extremely lightweight netting, made out of polyethylene threads interwoven with conductive stainless steel fibers keeps the chickens in and predators out. Inside these quarter-acre, 450-foot perimeter circles are two 20-foot hoophouses on skids. The nest boxes hang inside these hoop schooners. Moved every three days, these feathernets net about $10,000 per year at 7 person-hours per week on 5 acres. We graze with the cattle and eggmobiles around these units as necessary.

One of the latest innovations is a quarter acre vineyard/turkey range combination. Usually, the big cost in vineyards is ground maintenance. But we use turkeys to graze the grass, fertilize and debug. The vines shelter the turkeys and grow grapes. The synergistic net income generation of this unit should be about $4,000 per year. That's $16,000 per acre! Who says you need a lot of land to make money farming?

Each of these enterprises generates enough income to stand

alone. When dovetailed with the existing parent enterprise, the economic value increases. Sometimes these enterprises are called "holons." In his classic book Tree Crops, J. Russell Smith envisioned multi-tiered production patterns that could look like this: range poultry under bramble fruits under tall fruit-producing tall shrubs under apple trees under mulberry trees. The multi-story production captures far more available sunlight than any single one or two tiers. Even though the spacing on any one of the species is farther apart than an intensive monoculture of that species, the total production per acre is far more. Assume that the spacing is 40 percent of normal, add up four species at 40 percent and you have 160 percent.

This complementary enterprise principle works in every part of the world, in every climate and in any situation. It offers opportunities for every single child. Some children may not want to farm specifically, but they can operate other related holons. Examples:

- make ice cream on a dairy
- make ground flour from grain
- turn the grain into muffins
- make apple juice from apples
- install a miniature golf course in the old bank barn
- bed and breakfast to value add eggs, pork and grain products
- inspected kitchen to make vegetables into pot pies
- bandsaw mill
- woodworking shop to turn lumber into furniture
- use barn walls to display artwork and crafts in local juried show
- tan hides
- make gloves and moccasins from tanned hides
- cut woods trails and offer trail riding for horse owners

- add a catered picnic for the trail riders

This list could go on forever, because there is no end to the enterprises and value adding potential on any given farm. It's limited only by creativity and warm bodies to pull it all off. As the enterprise develops more holons, the income generation makes places for even more nonfarm enterprises. When the farm is supporting half a dozen families, then it can probably have a full time mechanic or shop person to fix and fabricate.

A full-time accountant could be added before too long. That would free up the other families to do their enterprise even better. And on special big-project days like chicken processing or hog killin' everybody can come together and help out. Many hands make light work. The evening we move all the pastured hens into the hoophouses we call friends and turn it into a 15-person party, moving 2,000 birds in two hours. This completely changes the task. It becomes a shindig instead of a dreaded job.

We take this complementary enterprise issue seriously, and have brainstormed additional holons even beyond our lifetimes. We've employed the children and now we're working on the grandchildren. In fact, right now we've identified about 20 full-time salaries from this little farm. Let me itemize them so you'll know I'm not just making this up. As I list them, I want you to just imagine what a productive farm this will be when all these enterprises are functioning:

- salad bar beef
- pigaerator pork
- pastured broilers
- pastured laying hens
- turkeys

- firewood and timber--forestry
- sawmill--lumber
- woodshop
- rabbits
- sheep
- orchard
- grapes
- marketing/promotion
- delivery van
- vegetables
- small fruits
- farm and ranch vacation cottage(s)
- aquaculture
- hoophouse vegetables
- mechanic/fabricator
- accountant/bookkeeper
- trail rides
- chef for info-tainment dinners
- inspected kitchen--pot pies, bakery, etc.
- greenhouse

Oops, this is more than 20. Guess we're working on more than grandkids now. But isn't that exciting? Can you imagine all that coming off one place? Talk about a place of ministry. Talk about a happening place. Talk about production. This kind of discussion gives me chill bumps. I want it so bad I can taste it. Where are the farmers with this kind of vision? While I can't sleep at night for all the opportunities on this little place, most farmers can't even get themselves employed full-time on the farm, let alone anyone else.

One final thought. You don't need to buy additional land to do a lot of these things. These can be piggybacked on an existing farm. This is a way in. And we desperately need a way in for young people, to bring new ideas and youthful energy into the countryside. Wherever you are on the scale, from lonely elderly farmer to passionate young person, an opportunity for symbiosis exists here for you.

Complementary enterprises fill the most critical gaps in our current agricultural paradigm. They answer the production issue, the labor issue, the generational transfer issue, and the opportunity issue. Although they represent a fundamentally different way of thinking for stodgy cattlemen and John Deere Jockey grain farmers, complementary enterprises offer regeneration for any and all farms. Let's get busy.

Complementary Enterprises

Chapter 22

Creating a Sense of Plenty

Creating a sense of plenty is fundamental in a multi-generational family friendly farm. If the children sense that the cornucopia is going to run out, they will begin looking elsewhere for new fruit.

The "worn out farm" designation, popularized during the frontier days and the dust bowl period in our nation's history has unfortunately come down to our own day alive and well. Now sustainable agriculture is in vogue, but Robert Rodale, before he died, preferred the term "regenerative." He rightly said it was not enough just to maintain what we had; what was really needed was to regenerate. Tying a knot at the end of the proverbial rope and hanging on may be sustainable, but it's not regenerative.

In order to truly feel like a place for them exists on the family farm, our children need to believe in the regenerative concept. There is this idea in our culture that we have precious few resources, that we are right at the brink of using everything up. I remember debating energy policy in high school and college days. We had hundreds of expert opinions that we would run out of oil by the mid 1980s.

Affirmative teams proposed policies to deal with this impending catastrophe. In the early issues of Mother Earth News I remember reading Paul Erhlich's dire predictions about the end of humanity if we lost some little bug. Now to you tree hugging environmentalists, I know this must sound like something straight from Rush Limbaugh. But you must admit that somewhere between Limbaugh and Erhlich lies a balance.

Joel Arthur Barker's classic book Paradigms contains one of my favorite lists of all time, a list of expert prophecy. I've collected these and many more over the years because they fascinate me. Here's Barker's list:

- 'The phonograph . . . is not of any commercial value.' *--Thomas Edison remarking on his own invention to his assistant Sam Insull, 1880.*
- 'Flight by machines heavier than air is unpractical and insignificant, if not utterly impossible.' *--Simon Newcomb, an astronomer of some note, 1902.*
- 'Sensible and responsible women do not want to vote.' *--Grover Cleveland, 1905*
- 'It is an idle dream to imagine that . . . automobiles will take the place of railways in the long distance movement of . . . passengers.' *--American Road Congress, 1913.*
- 'There is no likelihood man can ever tap the power of the atom.' *--Robert Millikan, Nobel Prize winner in physics, 1920.*
- '[Babe] Ruth made his big mistake when he gave up pitching.' *--Trisk Speaker, 1921.*
- 'Who the hell wants to hear actors talk?'*--Harry Warner, Warner Brothers Pictures, 1927.*
- 'I think there is a world market for about five computers.'

--Thomnas J. Watson, chairman of IBM, 1943.

- 'The odds are now that the United States will not be able to honor the 1970 manned-lunar-landing date set by Mr. Kennedy.' --*New Scientist, April 30, 1964.*
- 'There is no reason for any individual to have a computer in their home.' --*Ken Olsen, president of Digital Equipment Corporation, 1977.*

Aren't those hilarious? Anytime you hear a gloom and doomer start picking dates and getting really specific, just pull this little list out and read it. During the Y2K scare a list like this would have brought some sanity to folks. While I do believe we've raped and pillage our landscape, I also appreciate nature's capacity to heal. When your daughter falls and scrapes her knee, she comes running to Mommy as if the end of the world has happened. A wise parent will not fawn over the incident and make over it as if it's a terrible thing. Instead, a wise parent takes a look, cleans the scrape, applies a bandage if necessary, wipes the tears and gets the child up and going again. Parents who expect their children to just scream and scream over a little scratch are not doing themselves or the children any justice.

I grew up hearing that Americans were cutting down all the trees. Our county now has twice as many trees as it did during the Civil War. Now that our family has our own bandsaw mill, I've learned quite a lot about logs and lumber. When you look at the lumber in old houses and old barns, if you know anything about saw milling, you quickly realize how many trees it took to build a typical post and beam structure back then. Or a log cabin.

Now we are able to take all the left overs and make chip board, plywood, wafer board, cutting the wood waste phenomenally. I think it's still immoral to be building stick houses heated with petroleum or electricity when they could be heated with renewable techniques, but let's also give credit where it's due. Necessity is the mother of invention. Things aren't static. All of us have read stories about backyard mechanics who invented 70 mpg carburetors only to be bought out and shut down by an automaker.

While I would be the first to suggest farmers should live lightly on the land, reduce landfill material, reduce petroleum use and steward resources carefully, I would also suggest that the market resiliency and human ingenuity will kick in before we're all huddled around some sticks in a cave. At least for the foreseeable future.

All you have to do is fly over the world, and you realize the world is awash in resources. They may not be fairly allocated. They may be wasted. But enough sunshine strikes the planet to yield enough energy to keep us going for awhile. This is why I say I'm a Christian libertarian capitalist environmentalist.

While the farming community overall creates a feeling of plenty, and more often than not produces above and beyond the demand every single year, most individual farmers do not see the same abundance on their own operations. Much of the plenty being created today is due to globalism, not really increased productivity from the U.S. Once all the apple trees in the ground in the year 2000 begin bearing, China will be able to produce four times as many apples as the U.S. Anyone familiar with production around the world realizes that the commodity price is projected to continue dropping for some time as volume increases. Technology and machinery are increasing productivity in developing countries, who are finally able to pay for these items due to the economic boom from global markets.

Far Eastern factories assembling electronic parts or tennis shoes receive money from the West which makes these countries able to purchase agricultural technology and machinery. Of course, amidst this plenty, the nutritional quality continues to plummet. We are awash in food, but it's not fit to eat. This happens in other subject areas. We're awash in schools, but they're not fit to attend. We're awash in religion, but most of it won't do you any good. We're awash in music, but how much of it is wholesome? We're awash in Hollywood movies, but how many do you really want to see?

Meanwhile, back on the family friendly farm, we're looking at the landscape and wondering how it will produce more. As we step back and look at this question, I think a historical perspective from the top is insightful.

The Israelites left the bondage of Egypt and traveled across the wilderness into the land of Canaan. Before Joshua died, God laid out very specifically the boundaries for each of the Israelite tribes. They were even told, again very specifically, exactly how big the land of Israel was supposed to be. And while God gave them definite orders to take over, subdue and inhabit that land, He did not authorize them to extend their borders.

In the continuum of history, this was highly unusual. Most conquests never stop. When you think back on the Chaldeans, the Persians, the Greeks, the Romans, and whatever other empire you care to plug in here, you see a highly successful civilization that begins extending its boundaries. It gradually extends and extends until it overextends. Then it crumbles.

But the Hebrews were different. They never developed a thirst for conquest. They never looked covetously at neighboring nations. God gave them no sanction for additional conquest. In fact, He told them to stay put in this promised land. But the curious thing is that He told them numerous times to "be fruitful and multiply."

What is the lesson? The lesson is that under proper stewardship, the land should become more and more productive. It should support more and more people. Only under mismanagement does land degenerate. Anyone who thinks this illustration bears no import today, remember that all this occurred long after Eden's paradise.

We went down a few years ago to visit one of our former apprentices in the Mexican state of Sonora. Driving south out of Douglas, Arizona and heading into that desert high country from Agua Prieta toward Hermosillo, we were struck by the lack of vegetation. Hillside after hillside was totally barren. A goat would need 20 acres to survive. In fact, people who had cows were struggling to keep them fed at a ratio of nearly 100 acres per cow. No wonder coyotes killed many of them. You'd see a cow about once every half mile, just out there on the hillside trying to find a morsel.

Arizona and New Mexico were not much better. And yet I knew this area was famous for cattle production, buffalo and cowboys. Our apprentice's grandmother came to dinner and she began explaining what the landscape was like when she was young. From a wealthy family, she had attended a girl's prep school outside Mexico. She said it was difficult to actually see her native countryside when driving along the road--the same road we drove on--because the grass was so high you couldn't see out.

I was flabbergasted. This land that we had just driven through, where we could see for miles and miles, was a lush prairie. I could scarcely believe it. And yet my subsequent inquiries have proven its veracity. Of course, I've grown up on old farmland scarred by 16 foot deep gullied hillsides and truck-sized limestone outcroppings that 200 years ago were covered with 4 feet of soil. But this brittle environment in the southwestern U.S. and northern Mexico was far less forgiving.

In their fascinating book Climates of Hunger, Reid Bryson and Thomas Murray document and describe how civilizations have encouraged deserts. Greek historian Homer, author of The Iliad and The Odyssey, walked across northern Africa and never left the shade of a tree. Today it is the Sahara. Desertification is occurring in the U.S. as fast as in any country in the world. One of the most fascinating statements in Bryson and Murray's book is this one: "A fence built around a large field in the desert brought abundant wild grasses in two yeas, without planting or irrigation, simply by keeping out the men and goats."

Of course, all of us familiar with the work of Allan Savory and Holistic Management realize that it is not just keeping animals off the land that heals it. It is the timed control, the management access, that heals the land. In fact, as Savory would quickly point out, it is a fundamentally different way of thinking about the land, the grass, the animals and the tribe that creates the framework for healing.

The consuming burden on farmers that we live with limited resources, limited land area and therefore limited production and limited income potential could not be farther off base. Most farmers' experiences surely confirm this overriding thought, though. Farmers have watched their net income fall year by year. Their expenses rise. It takes more inputs to get the same kick. More vet bills. More disease. In fact, we lose more crops to insects now, as a percentage, than we did before pesticides were ever used.

While I appreciate the old Chinese proverb that if we keep going the way we're going we're going to end up where we're headed, I also put great faith in an individual's ability to turn around. Whether this can happen collectively in time to avoid some of the doom and gloom the environmentalists prophesy is anybody's guess. I am practical enough, however, to not fret over the direction of the entire culture because I can't do anything about it anyway, any more than I can make my neighbors heal their land instead of continuing to destroy their soils and the life within the soil.

But I can help individuals to understand that they can create a world of plenty on their own niche of God's creation. I realize none of us is an island, and if the whole country turns to desert, I probably can't keep my little spot green. That may be true. But if all I do is wring my hands and fret about all the people and places I can't change, it will suck down all my energy to turn my own place into a promised land.

We have watched our own farm go from scarcely supporting 10 cows to being able to easily handle 75. I've watched our little mudhole ponds produce some nice bass, and have every reason to believe that now that they are cleaned out they will produce far more fish. I've watched barren hillsides heal with verdant trees and shrubs. I can tell you that this farm is probably converting 10 times as much solar energy into biomass--energy--as it did 40 years ago.

It's a farm of plenty, not scarcity. When all the children hear are dire warnings about things running out, whether it's food, money or fuel, they feel opportunities for themselves slipping away. But when they grow up with more abundance every year, and a farming model predicated on plenty, they sense a multi-generational opportunity. A full pantry, a full freezer, and a full bank account. Creating a sense of plenty is one more critical element in the family friendly farm.

Family Development

Chapter 23

Greenhouse Kids

Family friendly farming offers a wonderful model for growing greenhouse kids. When our children were still small, I was routinely accused of raising kids in a sheltered environment.

"I want my kids to experience the world," many folks would say. "How do you expect them to ever function in reality if you never expose them to it?" they'd ask.

Drawing on my experience, I have developed a metaphor to answer these accusers that has now been borne out in our children, as well as countless others. We call ours greenhouse kids.

When I suggest that children's exposure to the outside should be limited, and that a fair amount of their time should be devoted to meaningful work, I know I must sound like someone straight out of Pluto to many people in today's western culture. Modern American childhood is crammed full of running here and there, Headstart and early Montessori schools. It's interactive museums and school programs.

In their seminal book Homegrown Kids, Raymond and Dorothy Moore address the kindergarten lure like this:

> *"Well, why not kindergarten? Here are some of our reasons why not: You must first decide your priorities. For example, few people would question the idea that physical health is more important than early education. The body is the battery for the mind, and your child's mental progress is largely dependent on his total physical fitness. So is the strength of his nervous system, his emotions, and even his moral stamina. School even for half a day at age five is not ideal from the health point of view. Until the child is eight or nine, he is especially susceptible to respiratory diseases. He needs to build up immunities to resist the germs to which he will be constantly exposed at school.*
>
> *"There are very few kindergartens that do not keep the children relatively confined in a classroom most of the time. Lack of physical exercise and the scarcity of pure air in the schoolroom are harmful to the lungs. Even with so-called modern methods of heating and cooling, many rooms admit little or no fresh air, especially in very cold or hot climates. The existing air is often recycled through the heater or cooler. In most homes there are relatively few people to take up the available oxygen which is vital to healthy lungs and maximum brain development. If the child has a long bus ride to and from school, the problem of excessive fatigue is added.*
>
> *"Reading-readiness workbooks and other near-vision academic work, as well as small-muscle activities such as printing, coloring, cutting, and pasting, are usually a large part of the programs. There are exceptions, of course; yet under normal circumstances there is no other way for the teacher to handle the number of children in her care than by a certain amount of regimentation and busywork activities.*

Her children must pretty well do the same thing at the same time or there would be chaos. Yet the children's eyes are usually damaged by too much close work, and the combination of pressures tends at early ages to make them nervous and high-strung. Manipulation of a pencil or crayon is still quite difficult at age five, especially for boys. The child may like to do such work at home, but there he is free to stop and rearrange his program at almost any moment, while at school he is more often expected to finish his assignment."

Sound amazing? It is. Add to this the findings of John Holt, the razor-sharp perceptions of John Taylor Gatto and countless other leaders in the homeschooling and unschooling movement, and you find a whole world questioning the conventional schooling mind set. But even within homeschooling, the greenhouse concept is too often overlooked.

In fact, perceptive homeschoolers differentiate between "homeschool" and "schooling at home." The Moores deal with this extensively in their sequel Homeschool Burnout. Rest assured that for every positive, wholesome paradigm, an unwholesome one rears its ugly head. When the neighborhood one-room schoolhouse worked great, along came age-segregated, consolidated factory schools. When homeschooling worked great, along came curricula producers and all the trappings of the factories. Some homeschooling parents even do bulletin boards and scheduled recess.

The idea of a greenhouse is to give the plants a head start. I realize in some applications, the plants stay inside their entire life. For example, January tomatoes in Alberta had better be in a greenhouse. But for the sake of discussion, I want to use the greenhouse idea in its most common application: a head start for plants.

If we plant seeds in harsh late winter conditions, the environment will wreak havoc on the little seedlings, assuming they germinate at all. Soil temperature, soil nutrients, soil texture, soil moisture all have to be right before the seed will successfully germinate. The longer it takes the seed to germinate, the less energy is left over for sending forth the shoot and roots.

If I came to your greenhouse in January and you were planting tomatoes in flats, you would look at me like I had two heads if I accused you of sheltering these plants. "Don't you know these seeds have to make it in the real world?" Both of us would inherently understand the value of getting those seeds off to a good start. You would carefully mix the potting ingredients and check the temperature with a thermometer. Each seed, carefully poked in with a popsicle stick, would be kept moist with misters or a plastic covering.

You would try not to overlook any component in the system that would get these seedlings up and running. Why? The whole goal of a greenhouse is to get plants up and running faster than their garden-planted counterparts. Gardening books are full of cloches, glass bells, floating row covers, and all sorts of protective films to insure a perfect environment for these plants.

The whole goal of keeping kids home and protecting them from the world early on is to give them strength and stability, to establish their stem and their viability early so they can take the elements and thrive later on. Any gardener knows the difference between two sets of plants, one started under ideal conditions and one started under harsh conditions. The greenhouse-protected plants will far outperform their outside-planted counterparts.

As the greenhouse plants begin growing, we gradually harden them off. Hardening off usually starts by dropping the temperature in the greenhouse. Letting the plants get chilly at night will help them acclimate to cold temperatures. Taking the plants outside during the day, or opening windows, allows the plants to adapt to wind. Anyone who has ever started seedlings on a windowsill knows what a spindly plant looks like.

And anyone who ever planted a spindly plant knows how poorly it performs. When I go to a greenhouse to buy cabbage sets, I look for a kink in the stem that indicates a spindly frame. The ones I buy may be short, but they're stout and rigid. These will be the ones that were appropriately hardened off, and they will continue to perform at peak ability when transplanted. The spindly ones fall over and, if all goes well, finally get going a couple of weeks later.

So another principle with greenhouse kids is to have a hardening off period. We don't apologize for starting them in a protective, sheltered environment. But at the same time we don't want to keep them sheltered their whole life. On a family friendly farm there's a fulness of time, a balance for all these elements.

A child's value system is not really in place until about eight years old. How many parents have sent their obedient, lively little tykes off to school and then spent the next 10 years trying to eradicate all the bad habits acquired from all the snotnosed kids across town? How many things do your 10-year-olds do that they picked up from other kids?

Children learn from the people they spend the most time with. This is what Rick Boyer, author of numerous home parenting books, describes as the difference between socialization and being sociable. When children spend all their time with other children, they are properly socialized, but who wants a household of little socialists? They are doing what everyone else wants to do.

I wish I had a nickel for every time someone has accosted me at a conference for not properly socializing my children. Honestly, I don't want them socialized. As Boyer points out, what we really want is for them to be sociable. And there's an important difference. Being sociable means having social skills, not social mimicry. Social skills are things like being able to hold polite conversations, thinking of others, eating gracefully and holding doors for ladies. Socialized kids take the path of all other kids.

And kids can be incredibly tough on each other. If you grew up with a physical defect, you know what I'm talking about. A sociable child would not point and make fun of your big ears or your facial mole. But socialized children taunt and make up nicknames about any anomaly they can find. When Teresa and I got married, we committed ourselves that we would not put our children through what we went through.

I was a good student, but I had zero values before about 10 years old. And I was a hellion. I got in fights, taunted, got beat up by bullies, and did all sorts of mischief I wish I could forget. And I was a good kid. But I got sucked up in the mentality of the day. I wanted to be liked, wanted to be elected, and was willing to do whatever it took to get there. I wasn't strong enough to walk away, to say no, to stand up to the temptations. And I had a wonderful home, good discipline and excellent parents.

Teresa and I knew that if this is the way it was for us, in that day when we still had devotions in public school, we couldn't expect our children to be immune to the same rigmarole. We were looking for an answer, but didn't know where to find it. We couldn't afford private school. And the kids we'd seen come out of private school by and large had the same attitude as kids in the government schools.

Daniel was about six weeks old when we were driving to a sustainable farming field day and turned on the car radio. It was in 1981, the first time Raymond and Dorothy Moore had been on James Dobson's

Focus on the Family. We didn't catch the introductions, but we heard the term homeschool. It was like a streak of lightning. I looked at Teresa and said: "There's the answer." I didn't even know what it meant, but it spoke to my soul at a moment of need, and I knew it was the right thing.

I think that's the way a lot of things are in life. That's why I don't spend a lot of time arguing with people. If I'm speaking at a conference and a heckler starts in on me, I usually just listen quietly, smile, and say: "Okay. It's a big world. You're entitled to your opinion. Next question?" If the things I've said in this chapter have you hopping mad, ready to throw this book out the window, I'm sorry. Spit it out and go on. Take what you want and throw out the rest. No problem.

But if this ministers to your soul, as a salve, as what the New Agers call the "Aha!" then you'll appreciate where I'm coming from and the greenhouse metaphor will help you answer the naysayers with positive forthrightness, rather than apologize. As a good kid who spent ample time in the principal's office, I can tell you that a deficient value system is just as dangerous as a deficient immune system.

How does this relate specifically to family friendly farming? Because today's children, even farm children, are not encouraging their peers to be farmers. And guidance counselors are even worse.

I can remember being castigated by the school guidance counselor when I wanted to take typing instead of physics. College-bound students were expected to take physics. But I planned to do at least some writing, if not a substantial amount. I watched these journalists hunting and pecking on typewriters at the newspaper office where I worked part-time in high school, and was struck by the inefficiency of it all. Today it's even worse. We have all these college-bound students hunting and pecking on multi-thousand dollar computers.

Greenhouse Kids

I judged a regional 4-H public speaking contest a couple of years ago and listened to one of the contestants give a speech about how anti-farm schools were. His main problem was the guidance counselor, who couldn't imagine why any student with brains would want to be a farmer.

Our entire culture stereotypes the poor, dumb redneck farmer to the extent that any young person with brains who wants to be a farmer has to mumble the dream and face public ridicule and embarrassment. Pick up any newspaper and you'll see the attitude carried along daily in Garfield, where the buffoon Jon Arbuckle is a farm boy, and Beetle Bailey, where the dumbest of the dumb, Zero, is the farm boy.

How many times, when I was winning 4-H poultry judging events, did I return to school and walk down the hall listening to a chorus of "Hey, chicken man" and a cacophony of "Awwwww, cluck, cluck, cluck." You bet I'm for greenhouse kids. The taunting only made me compromise my values more in order to be liked and accepted. If I hadn't been exposed to this so early, I think it would not have had such a big effect on me. But when we practice early exposure, we create a peer dependency that carries throughout life.

Our own children are completely different. They have not suffered the ridicule and temptation as little tykes because they've been learning social graces from Mom and Dad. Isn't it unfortunate that they won't have the memories of bloody noses, principal's offices, lying to parents and stealing gadgets to tell their children? How misfortunate. What a deprived childhood, that they'll only be able to talk about working, reading and playing with Mom and Dad in a wholesome environment?

Friends? Yes, they have some friends, but only a few. How many friends did we have in school, surrounded by hundreds of people? Only a few. But we've tried to make sure they just got gold friends, not fool's gold.

And now that their value system is established, we've hardened them off by being open and honest, incorporating them into our successes and failures, and they have their own sturdy value system in place. When I think of the anguish I must have caused Mom and Dad, and then listen to the anguish most teens cause their parents, I just bow my head in gratitude. Teresa and I are so blessed. But the bottom line is that it's not by accident. I don't think we've just been lucky.

We've devoted ourselves to our little plants. We got the soil right, put on the right kind of water, built a stout greenhouse, and got these little guys off to a roaring start. As I've said a million times, we're not really super parents. We've just practiced a very forgiving model. Modern society's early exposure model buffets the kids about with all kinds of fierce weather and hardship. It's a model with very little forgiveness. Just a little too cold, too wet, too dry, too hot or too windy, and the little plants can never perform like they could have and should have.

But the greenhouse model creates a huge buffer of forgiveness. It does require far more attention. It's not just a plant and leave mentality, but requires much preparation long before the seeds are even sown. The structure has to be in place, potting soil gathered up, flats made and cold-night supplemental heat installed. The comparison between the preparation for the greenhouse seedling and the outside garden-planted seedling shows just how much more the protected one requires. The difference is immense.

But the chances of success make the investment worthwhile. And when these kids get their day in the garden outside, they are head and shoulders above their companions. They are well proportioned, sturdy and ready to bear fruit. We can enjoy them more than their counterparts. We can expect a bumper crop. And we can enjoy their continued development, knowing they have achieved their fullest potential.

If this model draws you, and appeals to the yearning of your soul, what better place to do it than on a farm? A farm offers the seclusion and the protection that a greenhouse environment requires. Trying to duplicate this type of model in an urban condominium is much more difficult. The activities competing for your family time are more numerous. Besides, within the four walls only a certain number of interesting, meaningful avocations exist. Certainly plenty of good families and good kids exist in a metropolitan conventional schooling paradigm, but the batting average is far less.

Greenhouse kids. No apologies here. Every day I become more thankful for having raised greenhouse children. They are my joy and thrill. And when you meet them, you'll agree that the greenhouse model offers some real solutions in a crazy, zany world.

Chapter 24

Socialization: No Hermits Here

Most people in our society view farm life as a somewhat hermit experience. Stuck away from society, the family lives a sequestered existence. Occasional brushes with the outside world occur, but otherwise, life's social elements are nonexistent.

This idea has been cultivated by the farming subculture, whose members view city people as enemies. Most farmers want to be left alone. They view people as public nuisance number one. People gawk, crowd the roads and pass laws to put farmers out of business. Farmers see city people as lazy, materialistic and severely lacking in common sense. The recreational pressure just adds fuel to the fire. Cyclists. Big problem.

I went over to another farm we rent real early one Sunday morning and got behind a minivan loaded with bicycles. These folks were placing little cardboard arrows on road signs to show the day's bikers where to go. They were parked right in the middle of an intersection, so I stopped about 20 feet behind them. All of a sudden, I saw their back-up light come on and they began backing up. I couldn't get out of the way fast enough and these guys backed right into my car.

When the two vehicles collided, of course, the bicycle enthusiast jumped out of his car and apologized profusely. It was clear that he had no idea that someone else might be using the road. It was a beautiful clear Sunday morning in the country and he was here for recreation. It's all a pastoral landscape; animals take care of themselves; nobody uses these roads--they're just free for the taking.

I don't know how many times we've practically run over bicyclists on the blind turns around here. They are spread out five abreast coming around blind turns, completely oblivious that people actually live here and use these roads on a daily basis. And then they yell obscenities and give us the finger when we're swerving in the side ditches to try to keep from mowing them down. Yes, I confess there's a little bit of hermit in all of us who choose the country life. And the influx of recreational city people has created an emotional clash in our neck of the woods.

The mentality is further augmented by the notion among too many city folks that farmers are destroying all this beautiful farmscape. And it would be a lot better if the government owned it all. Environmental stewardship would be carried out for the public good that way, instead of the land being used and abused for the greedy self interest of these nasty farmers. If you're from the city, and you think I'm making this up, just come and talk to my neighbors for a little bit.

This clash between cultures was really obvious in the 2000 presidential election. The county-by-county colored maps revealed that the heavily populated areas went for the Democrat Al Gore and the rural areas went for Republican George W. Bush. The "DamnYankee" designation is alive and well in many areas, and has been broadened to include virtually anyone who comes from the city with a pocketful of money.

Read any farm publication, and you will see the opinion pages full of us vs. them diatribes. Farmers feel increasingly threatened by the politics of their metropolitan neighbors, and it makes all of us farmers want to run for the hills. But running for the hills doesn't solve anything, and it certainly doesn't impact the culture. It's important to see the silver lining in the cloud. These folks are potential customers, both for our farm products and for agri-tourism. Change always has its blessing and curse.

City folks see and hear the communication coming from the farm folk. The "head in the sand" mentality farmers portray is real. City folks believe they think and do what's best for the nation, and farmers are just withdrawing from their responsibilities for clean air, clean water and productive soil. Of course, I agree with much that the city folks say. But where we part company is in the solution. The solution is not more government programs and regulations. All of this makes the farmer feel hemmed in, isolated and not a part of the big, swirling world that's out to get him.

All of this comes together to make would-be farmers fear that a country experience is a hermit experience. Isolated, withdrawn, socially deprived and culturally stunted children will be the result. That's the fear. And then when you throw in homeschooling too, the existence seems unbearable.

Let me tell you something: We're not hermits. Take heart. The family friendly farm, while being far from big city lights, is far from a hermit experience.

The family friendly farm is magnetic. Because of its diversity in production, it's an interesting place. And interesting places are always fascinating to visit. How many people want to visit a 1,000 acre corn farm? Nothing really exciting happens, except during plowing and combining. But a few minutes of that, and you've seen all there is to see. It's the same thing, over and over and over again.

It's more interesting watching excavation for a housing development. At least more things are going on at one time. Most farms are simply places where a couple of times a year big machines come out and rip up and down the landscape. Between times, everything just sits. And you sure don't want to visit during chemical application times.

Because we direct market, a steady stream of customers comes out to the farm. The children show these folks around, interact with them, and play with their children. Many of these customers are friends as well, and when they arrive to purchase things, their children and ours play games or do whatever kids do. The visit offers both social and business elements. For the adults, the same holds true.

Unfortunately, farming has become so family unfriendly that zoning laws now imperil much of what we do. Our county now takes the position that direct marketing falls under retail sales, a prohibited use in agriculturally zoned land. Our county told a local sheep producer that if he sent his lambskins off for tanning, and then brought them back to the farm for sale, those were manufactured products and illegal to sell at the farm.

Using the same reasoning, if we send a pig off to be processed, bring the frozen, labeled packages back, and then sell the packages here at the farm, we're selling a manufactured product. The county argued that any raw product which leaves the farmgate for value adding or further processing, and then is brought back to the farm for sale, is a retail, manufactured item and cannot be sold in agriculturally zoned areas.

A friend down the road wanted to have a small, commercial woodworking shop next to his house. But because he is in an agriculturally zoned area, woodworking shops are prohibited. Several years ago I fought just to maintain the right for us to mill lumber from our own trees. The county tried to zone sawmills out of agricultural areas. All of this is done because sawmills, tanneries, cabinet manufacturers and the like are generally huge factories that make a lot of noise, create large amounts of traffic, hire many employees and create noxious fumes.

We have been desperately trying to salvage enough freedom to be able to sell our own vegetables, animals and lumber, but it is a real uphill battle. Since no one wants manufacturing in their backyard, they've zoned it out of everything except special designated areas. But like all the rest of these compartmentalized solutions, all the little guys get caught in the melee. Imagine my chagrin the other day when a customer was here buying some things and we began talking about people moving into the area and subdividing farmland. Her solution?--"We need more zoning." I had to restrain myself, believe me.

I wanted to scream: "We don't need any more regulations. If we did away with the ones we have, it would end corporate welfare, thereby creating a level competitive playing field between big guys and little guys. Then if we little guys could access the market, farmers would make enough money so they wouldn't have to sell out to developers. The regulations impede farm income, creating additional and unnecessary hurdles against economic viability. We don't need subsidies, we don't need programs. We haven't had a truly free market for a hundred years in this country."

People are incredibly naive. They think that if we would all just care, everything would be fine. If we would just love the land, then we could have policies to usher in the millennium. Forget it. "But don't you understand we must plan" say the sincere think-gooders. The problem is that your plan will be my undoing. Salvation by legislation never works because by the time the sincere-sounding idea makes it through the lobby merry-go-round, through the insider machinery, through the bureaucracy fraternity, and finally gets implemented in the countryside, it doesn't look anything like the critter first envisioned.

The endangered species act has destroyed incredible amounts of forest acreages as loggers try to harvest timber they never would have harvested had they not feared it would be locked up forever in some critter protection program. The no-clear-cut policy means that harvesting now is far less efficient than before. Now loggers do a select cut, which damages all the residual trees. Two years later they come in and do a salvage cut, which in effect is a clear-cut spread over two or three years, but it cost twice as much money to come to the site twice instead of once.

The result is more disease, lost production, higher priced lumber and more taxpayer subsidized timber sales. The problem is that we have thousands, even millions, of people who feed on a media agenda by a narrow interest activist group, and who have never owned any land. It's always easy to tell someone else what to do, especially when we've never done it or experienced it. I'm not for huge clear-cuts, and I certainly love wildlife. But somewhere between the extremes lies a balance, and that balance is best determined by informed customer choices. Informed customers occur when big brother government allows freedom and diversity to proliferate.

As long as everyone thinks big brother government will protect them from ever making a wrong decision, individuals will gladly let the government do it. And so the clash between special interest groups--farmers vs. environmentalists, etc.--continues to escalate, creating more ignorance and deeper compartmentalization. The parceling out of product, land use and how society operates creates additional tunnel vision on the part of the citizens, and pushes all of us toward hermits, living in our little world.

But our own farm customers provide a constant sharpening for our social skills. They also give us a deep appreciation for the urban agenda. These are two-way conversations, and help us to live balanced within the social framework. The average farmer has never had a meaningful conversation with a metropolitan environmentalist. These folks comprise probably half of our patron base, and create in us an appreciation for their thinking. It also produces accountability in our thinking, planning and action. It is always good to be touched by people who hold different viewpoints. Evolutionists and creationists should read each other's books. The same holds true for protectionists and globalists, public schoolers and home schoolers, libertarians and consumer protection advocates.

Of course, our church family provides plenty of social activities, as well as our home school support group. The numerous activities available mean we do not avail ourselves of all the possibilities. Our children have their so-so friends, close friends and distant friends. Rachel is our letter writer, and corresponds with about a dozen girlfriends from around the country. Some are children of family friends, and others developed as a result of a farm visit, or even a farm conference we attended. Given any opportunity at all, children will develop friendships.

Because our children are sociable, we have never considered them a liability when going to more formal social gatherings. As a result, we have virtually taken our children with us everywhere, including places where no one else would bring children. Obviously, if they aren't invited, that's one thing. But otherwise, we're glad to bring them along and let them share the experience. Why should they hear about it secondhand?

The trouble is most parents consider their children a bigger social liability than the family dog. In fact, many pets are better trained than children. As a result, parents dare not expose their children to public social shindigs, where they may cause an embarrassing scene. Properly disciplined, well-adjusted kids are a joy, especially to senior citizens. Christmas caroling in nursing homes makes both a child and an older person come alive with the experience.

Because we do view public speaking and public presentations to be a weakness in homeschooling, we've enjoyed Daniel's participation in 4-H contests. Although he's naturally gifted in this area, these competition opportunities have given him increased skill and poise, as well as leadership and many social opportunities. He has learned many mixer, get-acquainted type games and now uses them in other situations. Challenge courses, used for team building on the state 4-H cabinet, have been incredibly beneficial.

The apprentices, who rotate annually, offer many new social

experiences. The relationship is long-term and intimate, necessitating ongoing communication, forgiveness and emotional tolerance. Some families participate in Fresh Air programs to both minister beyond the family and allow the family to be touched by outsiders. This can be socially healthy, and should be cultivated.

For tax purposes, Teresa keeps a record of business meals she prepares for visitors, and the count usually runs about 500 per year (almost 1000 during the year 2000). That does not include entertaining friends. That's just business hospitality. Does that sound like enough social activity?

The bottom line is that not only do we have plenty of social elements in our lives, we have to be careful that it not dominate everything else. Anyone who likes people will have no problem maintaining a social life. The problem is letting it get so hectic that it destroys the family's sanctity. The challenge is not in getting enough; the challenge is in getting too much.

Farming does not require a hermit's existence. Nor indeed should it. A family friendly farm has a healthy balance of private time and social time, producing confident, passionate kids with beamingly proud parents.

Socialization: No Hermits Here

Chapter 25

Baggage: Dealing With It

All of us carry baggage. You have it and I have it. The big question is this: "What are we going to do about it?"

First, just so we're on the same page, let me describe a little bit of baggage. I think the metaphor helps because it's a perfect word picture of what's going on. The worst thing we can do is deny that we're carrying it.

Baggage is all that emotional pain and dysfunction that parents bring with them from childhood. And in a multi-generational family business, the baggage gets compounded with in-laws, step-relatives and so on. Shared vision and shared experiences are less common in the extended family situation. The baggage compounds from generation to generation unless we do something fundamentally different. Some of the baggage is obvious, but much of it is not. To jog our thinking, here are some possibilities:

- Growing up in a home of conditional love. As a child, did

you ever feel like Mom or Dad loved you only when you brought home good grades, or when you made it to first base, or when you got the starring role in the play? Did you ever think your parents would love you more if you performed better?

- Suffering the emotional pain of divorce. Watching the two people responsible for your spiritual, mental and emotional health, the two people whose passion brought you into the world, part company leaves lifelong pain and anguish. An unstable home life denies children good patterns, good models for their subsequent marriages.

- Feeling like you're a burden to the family. I have been horrified to meet people who never heard their parents say "I love you." Or worse, whose parents ignored or said things to them like: "You were an accident. We didn't want you. You got in the way of our career plans." I think I could hang parents like that. I can't imagine that they exist, but I know they do, because I've talked to their kids.

- Alcoholism or drug abuse. The pain caused by substance abuse endures a lifetime. Seeing a parent in a drunken rage or babbling like a baby creates emotional scars that insure dysfunction for years to come. Often children in these families develop a thick emotional shell, and block out the hurt with a spirit of autonomy that makes it hard to express warmth and emotional attachment.

- Meddlesome parents. Did you feel like you could never do anything on your own? Controlling parents, micro-managing your life, can leave you timid and unsure of any decision you make.

- Promiscuity. For a young woman, the feeling that all I'm good for is my body. For a man, a disrespect for women and a smugness about his dominance and "hunting" ability. These emotional scars often last a lifetime and affect marriages, sometimes forever.

- Jeering from other kids. This could be anything from having some slight physical defect to being overly quiet. Late bloomers catch this in middle school. The feeling of inferiority manifests itself in all sorts of ways--often in a "hunter's instinct" once puberty is over.

- Latch key children. Unsupervised siblings can get into all sorts of trouble between themselves. The lies, the coverups before Mom or Dad gets home. My! The stories parents find out about 30 years after the fact.

- Undisciplined childhoods. Did your parents ever say: "You'd better move before I count to three?" If they did, you've been abused. You grew up thinking rules were meant to be bent, and words didn't really mean anything. "If I have to come up there one more time!" really means: "I didn't mean what I said awhile ago, and you can count on me to be frustrated by you. You are in control, I'm not. I don't have any answers."

- Overly disciplined childhoods. "You'll just have to sit there all night if you don't eat those beans." Mealtime should be a happy time, not a tug-of-war. We've watched children vomit in their plate while being forced to eat "that stuff." It's not worth it. How many things do we adults enjoy eating now that we didn't as children? I couldn't get within 10 feet of Swiss cheese without feeling nauseous. Now Teresa gets it for me as a treat. Fortunately, nobody made me eat Swiss cheese growing up, so it didn't have any additional bad experiences associated with it when I finally took a bite as an adult.

- Unresolved parent/child conflict. I have not always spoken perfectly to my children. Has any parent? But these hot times need to be cooled off sooner rather than later. All this dirty stuff needs to be resolved, or it just builds up like gunpowder and gets released all at once in inappropriate ways.

- Unresolved sibling conflict. Things get said among family members that can never be forgotten. These things get carried over into adulthood and linger just below the surface at Thanksgiving, Christmas, birthdays. If it's serious, it can develop to the point that, were we not related by blood, we would have nothing in common and no reason to ever see each other.

- In-law, step-parent conflict. Only if you have been through the pain does this one touch you deeply. The wedding day that got taken over by an aggressive in-law; the grandchild that should have been a brunette instead of a redhead. Oh, how these words cut.

- Disinterested parents. This one is devastating. Children who never feel like their parents are looking at them when they speak. Preoccupied parents, always immersed in their world with no time for the child. A fellow told me he was in his second year of college and his dad asked him if he'd finished high school yet. This was not a joke--it really happens. Stories abound about parents trying to buy affection from their children with toys and money instead of time.

Tough stuff, isn't it? When I read a list like this, I just shake my head and smile as I go through the ones that don't apply, but then, sure enough, there's mine, and I feel this big knot well up in my throat. My eyes water, and that deep pain rears its ugly head and I relive that incident all over again. Oh, why won't it go away? Why, why why?

This is baggage. Everybody has it, and it jaundices our perspective. It keeps us from being able to communicate. Let me relate a story to show just how this happens.

We rent a small pasture farm about 10 miles away that the current owner bought at an estate sale. The previous owners, as they aged, deeded off lots to their children. Now the farm has almost no road frontage. Some of these folks bear a bit of resentment that the farm passed out of the family. A 50-foot tip extends down between a couple of these lots to the public road. That was preserved for an access to allow further lot development.

We put a two-strand electric fence down around this tip and the field to keep the cattle in. This fence doubled as the yard fence for an adjoining house and lot, now owned by a young couple. The husband's mother, a direct descendant of the earlier farm owners, lives two lots up, and this lot also adjoins the farm entrance. We had grazed this field tip for a season and were getting ready to move the cattle into it during the first cycle of a new grazing year when Grandma came marching out, arms akimbo, and across her back fence screamed: "You can't put those cows down there!"

I couldn't have been more blind-sided if lightning struck me

out of clear blue sky. I was about 40 yards away, moving the calves into the last paddock before entering this tip field, and what she said didn't register right away. I looked up and she screamed again. After moving the calves,
I walked over to where she had planted herself in the yard and asked her what she meant. She repeated the command: "You can't put those cows down in that field."

I asked her why, and she said her granddaughter lived down there. I had seen this little girl playing with her parents in the yard the previous year. "If she hits that electric fence, it will hurt her," explained Grandma. I toyed with the correct response.

Finally, I said: "Oh, it might scare her, but it won't physically hurt her. You see, this is a special pulsing charger. It's not like sticking your fingers in a light socket. It gives a sharp jolt, but doesn't do any irreparable harm."

"If she hits it, she will die," announced Grandma.

"Ma'am, I've hit this wire. My kids have touched this fence. It won't kill anybody, believe me. Besides, if you're that concerned, just keep her away from it," I said.

"You can't tell a child not to do things," she replied.

At this point, I was trying to keep up with what was NOT being said, and it sounded to me like this was a case of overprotective Grandma and spoiled little brat who needed some discipline. What do you mean you can't tell a child what to do? I felt my indignation beginning to rise. After all, I grew up in a home where spankings reminded us of where boundaries lay.

"Ma'am, we had those cows down there last summer and everything was fine. What is different?"

"Last summer, she was too small to get to it. Every time she went outside, her parents were right with her. Now she is walking and can go anywhere."

That made sense. So I decided on a compromise: "I'll tell you what. We only have the cows in there about six days in the whole year. That's the only time we have spark on the fence. Suppose I call your grandson and tell him when we're going to put the cows in there, and they can either keep their daughter in on those days, of if she wants to go outside, be sure to go out with her. It's only six days a year."

"You can't tell a child what to do," she said, seemingly incredulous that some parents actually have children that listen to their commands and obey. At this point, I reverted to the "spoiled brat" notion.

"Ma'am, we believe children can be taught to obey. Look, if she touches it once, she may cry, but she won't touch it again. Have you ever heard of spanking?"

"Listen, young man, I am not going to stand here and let you kill my granddaughter. That fence is illegal. It is illegal to have an electric fence around a yard. I'm calling the authorities."

And with that, she stomped back to the house. Her big dogs held their ground, though, barking at me ferociously until I was halfway back across the field. So much for diplomacy. The cattle were due to graze there in two days. I decided to make a couple of phone calls.

The first call was to the county magistrate. I explained the situation and asked him if the fence was legal. He confirmed my hunch that no such law existed. Legally, I was okay, but everything that is strictly legal is not right. I wanted to be a good neighbor, but obviously I was not getting anywhere. By this time, I had decided that this was just a family that didn't believe in disciplining children, and perhaps they were a little resentful that I had access to this ancestor's field.

The magistrate told me that we could be in a gray area since this tip bordered a public road. That had to be a "legal fence." For those of you unfamiliar with fencing laws, a "legal fence" is defined as one that will hold cows. How do you know if it is not a legal fence? If cows got out, it is not a legal fence. Circuitous reasoning at it's best. I decided we probably ought to build a good woven wire fence, but I was just two days until grazing.

One other part - to help keep the story on track. During the previous winter, between grazing seasons, this young couple approached the landowner about buying that field. The landowner called me and asked if it was a good idea. I told him that it was the only piece of ground on his farm that lay correctly and was fertile enough to be a good tree fruit site. If he ever had a desire to grow apples or plums, that area was the best. It also had this access if he ever wanted to plat off a piece of property. End of discussion. I just wanted the landowner to know what that piece was good for. He thanked me and we never mentioned it again, except I knew that he called the neighbor and turned him down.

I decided to go visit the young couple with the granddaughter. When I arrived at the house, the mother of the girl was sitting on the front porch, and immediately accosted me as a child hater. She was loaded for bear, and I was big and ugly. She let me have it with both barrels. I can't even remember all the things she said--probably a good thing--but I didn't get many words in. It was a fairly one-sided conversation. I put two and two together, figuring that these folks were just trying to run us off the place because they wanted to buy it and the landowner and I were in cahoots, trying to dash their dreams. Vengeance kind of deal.

In a few minutes, the little girl's daddy came home. He was laid back, easy going. His wife stalked into the house and he and I were alone, leaning up against his pickup truck. I apologized for opening a firestorm, and assured him that I would build a non-electric fence, but I couldn't do it immediately.

Right away, he explained the whole deal: "Our daughter has a heart condition that kicks in under stress or any foreign shock to her system. The doctors have told us to protect her from anything that would shock or stress her because she could easily go into cardiac arrest."

Suddenly, it all became clear. Grandma was absolutely correct to be concerned about her granddaughter. I completely misjudged her. Why she didn't explain the condition, I'll never know. Maybe she thought I would gossip about her. But she was protecting a vulnerable granddaughter, and rightly so. The mother was doing the same thing, and reasoned I was up to no good because I wasn't backing down. I misjudged their motive as vengeful for not being able to get access to the place. I also wrongly assumed they had a spoiled brat. They thought I was incredibly evil for not caring about their problem. And they hadn't bothered to tell me about the fencing problem because they thought the owner was going to plant apple trees on the hillside and our cattle would never be in there again. All in all, we had conversations and assumptions that went right past each other.

A few days later Teresa and I took a turkey to the young couple as a peace offering and we all laughed and put it behind us. We grazed the field twice without electricity, but then stopped for the rest of the season after a couple calves got out. The next spring we built a brand-new, first class, "legal fence" on the mutually-agreed-upon line, and we've all lived happily ever after.

I don't know why so much emotional energy had to be expended before the core truth finally came out, but it seems like that is often the case. We wrestle with each other and wonder why that person acts the way he does, or responds the way he does, and most of the time it's because we don't know where the other person has been. What he's been taught, experiences he's had, disappointments he's been dealt--these things affect what we hear. When Teresa and I have dealt with couples going through separation we're always incredulous that two people can describe the same event so differently. To one the party was a smashing success, a highlight of the year. To the other, it was the worst nightmare and the most Devilish contrivance ever conceived. How can this encourage a family friendly farm?

We bring baggage to our marriages. We bring dysfunction from our childhood and foist them on our children. Imagine a group of people sitting around a room with their baggage. Think about your family for a minute. Just imagine you're all waiting at an airport check-in line. Mom has this baggage; Dad has his; sister's looks like this and brother's looks like that. I think it helps us all to visualize each other with our bags. Different shapes and sizes. Some of it is old and worn. Some of it is brand-new stuff. Some is so huge we turn around and say: "What in the world do you have in there? Why did you bring all that along?"

Sometimes we have so much that in comparison it seems like everyone else is just carrying a briefcase. Or maybe we're so overwhelmed with ours, we can't see over it to see what anybody else has. "Why do I have all this stuff? You don't appreciate me because you don't have to carry all this. If you had to carry all this, you'd be more kind. You'd treat me nicer."

Let me tell you a secret. We all have baggage. The question is: "What am I going to do with my baggage?" I think every person's goal needs to be to work through it and throw out the unnecessary stuff. You'll never be baggage free, but it sure doesn't help any to hoard it, to let it be
the excuse for dysfunction. It must be gone through, honestly, piece by piece. And that is what we're here for, for each other.

I'm here to help Teresa live baggage free, and she's here to listen and work through my baggage. Your friends are part of this support network too. All of us need to understand we're dealing with each other's baggage. That should give us a little more compassion, but it should also give us a mission. While I need to be sensitive to your baggage, I'm not going to help you by letting you hoard it. Part of my job is to help you work through it.

And I expect you--my spouse, my friends, my kids, my adult siblings, my parents-- to help me work through mine. We hate our baggage, but we seem to cling to it. It's great to hide behind, to justify my feelings. It's awful when it keeps us from being who we ought to be. At first, being separated from our baggage can be threatening and create insecurity. But when we can walk freely about without dragging all this baggage, the working-through creates freedom.

How do we deal with family members who want to hoard

their baggage? I wish I had an answer. Ultimately, each of us must take responsibility for how we've dealt with our baggage. I'm not suggesting we should let any Tom, Dick and Harry peer into our most private bag, but generally I think we should be far more open than we are. Clamming up, taking my baggage, and going home is not the way to resolve anything. The older we get, the more baggage we accumulate and the heavier it gets.

In a family friendly farm, baggage must be fairly open. If you can't have open conversations about each other's baggage, better not get too involved. Intimacy is risky. Most of us have been hurt by it at one time or another. But what's the alternative? I'll tell you what the alternative is. The alternative is pulling into my shell, doing my own thing, not listening to others' advice, and becoming autonomous.

That's an intensely lonely existence. When we've been burned a couple of times we don't want to put our hands in the fire. That's only natural. But the long-term answer is not withdrawal; it is exposure and working through. If not with the people currently in your life, new people.

Our level of intimacy is only as deep as the openness of our baggage. If I can't trust you with what you see when I open my bags, I'll never let you see inside. There might be some things in there I'm embarrassed about. There might be some things in there that make you blanch. But if I think you're going to quit being my friend, run off, gossip about me, kick me out, then I'll never let you see in it. And we can never go to the next level of intimacy.

I know for some people opening that bag is the most painful thing in the world. But holding onto it, sleeping with it, taking it to every room in the house, is not the way to get free from it. I desperately yearn to help my family, and extended family, work through their baggage. And I want them to help me work through mine. But this desire cannot be forced. To work, it must be mutual. I don't have all the answers, but I do know that this issue surfaces time and again in family operations.

Perhaps it is helpful just to bring it to the surface. Just to bring to light this most private element in relationships. Because at its most fundamental level, family friendly farming is about relationships. Building them, healing them . . . and sometimes, cutting them loose. It's about dealing with baggage. My goal is to go through life with no check-on baggage. That lets the most people on the airplane, and lets us get there the fastest.

As our family sits around the table, here are some keys to working through our baggage:

- Mutually agreeable vulnerability and openness
- Recognition that all of us have baggage
- Radical honesty
- Assume no one is vindictive and everyone is compassionate
- Accountability--attitude, action, philosophy
- Some people don't want to cooperate--don't blame yourself for it
- Don't wait for someone else to open up first--go ahead and start
- Time is real--both to accumulate and to heal

Notice, I did not say: "Be loving." The reason is that if I'm holding you accountable, you may not think that's love. Love is too subjective. It's the easiest cop-out. "I'm not going to listen to you because you don't love me. If you loved me, you wouldn't say that." It's the classic cop-out.

I'm not going there.

All of us have a contribution to make when helping each other with our baggage. But no one has a magic wand to wave over humanity that will make us all deal with our baggage. Within the family friendly farm, however, we'd better all be making a good faith effort to deal with our baggage. Otherwise it will consume us. God help us to be courageous and successful as we work through our baggage.

Chapter 26

Noble Literature

"Where did you learn all this? Did you go to agriculture college?"

I wish I had a nickel for every time someone has asked me that at the end of a conference presentation. I have a stock answer: "If I'd gotten an ag degree, I'd be an egghead like the rest of them. No, I got it from Grandpa, Dad, reading, and visiting successful farmers."

In my own case, acquiring the information and techniques to be a successful, full-time, small farmer came primarily from the books and magazines that flowed through our home. Dad set the standard for what was worthy of study, and made this material the topic of conversation in the home.

Growing up without television--and still not having one--fostered conversational recreation, the incubator for the world of spoken and written ideas. Perhaps nothing stimulates animated conversation more than a mutually read piece of exciting information. Dad and I would digest these discoveries and debate their best application to our farm and its unique problems.

In retrospect, my love for reading what was right was actually more critical than my love for reading. And it was Dad and Mom who created the standards for the material that flowed into our reading pool. When I visit most farmers, I see conventional magazines and extension service bulletins lying on the desk or stashed in the toilet-side rack. I hardly regard these publications as promoting family friendly farming.

Being on the inside of the organic certification hoopla, I became an early dissenter of the process. I realized early on that food integrity could not be legislated or bureaucratically regulated. Farmers flocked to this premium-priced niche who didn't have an understanding or a conviction about the processes involved. And now the corporate empire is joining the organic movement, adulterating it further. In my discussions with the movers and shakers within the movement, I developed a radically different criteria for knowing if a farmer was really who he said he was: "Show me his bookshelf."

What a man reads tells a lot about who he is and what he believes. If a farmer says he loves the earthworms and uses no medications on his animals, but has a bookshelf and magazine rack full of nothing but conventional neanderthal chemical agriculture advice, I have serious doubts that he's as clean as he purports. But if I see literature about composting, mineralization, pasture-based systems and interspecies symbiosis, then I'll figure he's on the up-and-up.

Let me say again, probably the single greatest debt I owe Dad and Mom is not just a passion for reading, but the type of literature I read. When I see my counterparts still reading the pablum from the agri-industrial conventional establishment, I realize how the reading direction is more important than the act itself. The type of material is what insures that I keep discovering things that are valuable. We can discover all sorts of things through reading, but if they aren't valuable, aren't helpful, we've just wasted time.

While my counterparts in high school were familiarizing themselves with pop culture, Hollywood, sex rags and romance novels, I was reading *Mother Earth News*, the *Freeman* and Acres USA Primer. Not to mention Back to the Bible Broadcast material and memorizing Bible verses. And I realize now just how important that particular literature was in incubating lifelong vision, direction and passion that set me on a different voyage vocationally than most others. It also put me on a different search path. I continued to search for information that would complement the principles Dad had established. Vice-versa, my young counterparts sought for information that would shock their parents.

Rather than undermining my family value system, what I read confirmed it in my own mind and then to my own young family. Educators from John Dewey on have recognized the power of education, especially on children, to change culture. I would suggest that the big change agent is not the nebulous world of education, but the specific and practical world of what children read. Perhaps it's that simple.

Notice what rebellious kids read. They aren't reading what Mom and Dad recommend. They're reading counter-parental stuff, whether its chosen for them by teachers or whether they choose it themselves. The diet is counter-parental. But show me a child who loves to read the material recommended by Mom and Dad, and you'll find a happy, obedient and mature young person. Perhaps exceptions to the rule exist, but I haven't seen one recently.

Teresa, my helpmeet, has joined my love affair with good books by scouring and procuring anything pre-1950 agriculture for me. The folk wisdom and practical on-farm design of this era literally comes alive on the wall of our home office. Other books I enjoy at the moment are futuristic and trend-type books, which explain where our culture is heading. Leadership and business books, Christian political, as well as unconventional spiritual books, round out my current repertoire.

Few things please me more than to see Daniel and the apprentices engrossed in these classics that provided my philosophical foundations a couple of decades ago. Rachel has practically memorized the Little House on the Prairie books, and I'm confident this has stimulated her domestic and hospitality skills.

I can still remember sitting on Dad's lap while he read about the one-eyed Cyclops. During our down time in the winter we read all the classics to our children that Teresa and I never read in school because we were too busy reading snatches in modern literature books. To actually read aloud the unabridged great classics has been refreshing for us, giving us the childhood experience we missed.

Good reading material is probably one of the best financial investments we can ever make. It is a rare book indeed that doesn't give me at least a $30 idea. Few things yield the payback of a good book economically and emotionally. Investing in good literature yields returns multi-generationally, and that's a valuable portfolio for the family friendly farm.

Chapter 27

Balancing Stimuli: You Can't Do Everything

"How do you go to the interactive museum, the local band concerts, involve your kids in soccer and ballet, yet keep them interested in the farm?" The lady who asked this question at a recent seminar was speaking for thousands of parents.

The question has two underlying notions. The first is that children must be heavily exposed to everything all the other children are exposed to in order to be relevant. A hidden idea is the cultural fear that farming is necessarily archaic, reserved for rednecks and simpletons.

Unfortunately, that is the real mind set in our culture and is bolstered by every visit to the local sale barn. You'll see gallon cans every half a dozen seats to serve all the farmers as tobacco juice spittoons. Farming is definitely a subculture.

The old pendulum swings again between the extremes. It seems like if our child is a farmer, we also want him to be a successful rocket scientist, just to give us credibility and make the neighbors proud of our kids. Our society is not willing to accept a middle ground concept, that a farmer need not be a country bumpkin. And that our child need not be a rocket scientist to make the neighbors proud. Really this question exposes the supposition that if our children devote their lives to digging in the dirt, it reflects on us as parents. We've failed our children if they have no more gumption, no more cultural savvy, than to become farmers.

In fact, I am disheartened that even the homeschooling movement doesn't recognize farming as a viable vocation. The college for homeschoolers being put together in Northern Virginia is devoted to public policy--only one major is offered. I would like to know where George Washington, Thomas Jefferson and Benjamin Franklin acquired their common sense approach to life? For that matter, why did Jesus teach with parables, almost all of which were based on agrarian circumstances?

The answer is not that the Jewish culture was agrarian. Yes, it was heavily agrarian, but God had situated it strategically between northern Africa, with its rich Ethiopian and Egyptian cultures, and the powerful eastern cultures of today's Iran, Iraq and surrounding areas. Israel controlled the King's Highway, the main route of commerce between these great civilizations. The plan was for Israel to influence the world.

A life removed from the soil is one that quickly loses touch with reality. And a culture with no agrarian context becomes arrogant in its cleverness, loses its humility. The attitude of the men who signed the Declaration of Independence was not one of radical arrogance or knee-jerk activism. Firmly planted in business and farming, these men carried an uncommon sense. They debated with humility as they sought workable solutions.

Nurturing life, tending animals, listening to the birds sing in the stillness of a rural morning, gives a sense that we are very small. That reality is much bigger than we can imagine. That forces far beyond human control are at work in our world. The black eye on these men was their lack of respect toward the creation. Their dominance theme carried the pioneers onward to new land after they abandoned worn out farms. But this environmental arrogance, even in that day, was not a universally held belief.

John Taylor, a contemporary and fellow legislator with Thomas Jefferson, wrote in an essay around 1800:

> *"Let us boldly face the fact. Our country is nearly ruined.*
>
> *We have certainly drawn out of the earth three fourths of the vegetable matter it contained, within reach of the plough. Vegetable matter is its only vehicle for conveying food to us. If we suck our mother to death we must die ourselves. Though she is reduced to a skeleton, let us not despair. She is indulgent, and if we return to the duties revealed by the consequences of their infraction, to be prescribed by God, and demonstrated by the same consequences to comport with our interest, she will yet yield us milk.*
>
> *We must restore to the earth its vegetable matter, before it can restore to us its bountiful crops. In three or four years, as well as I remember, the willow drew from the atmosphere, and bestowed two hundred weight of vegetable matter, on two hundred weight of earth, exclusive of the*

leaves it shed each year. Had it been cut up and used as a manure, how vastly would it have enriched the two hundred weight of earth it grew on? The fact demonstrates that by the use of vegetables, we may collect manure from the atmosphere, with a rapidity, and in an abundance, far exceeding that of which we have robbed the earth. And it is a fact of high encouragement; for though it would be our interest, and conducive to our happiness, to retrace our steps, should it even take us two hundred years to recover the state of fertility found here by the first emigrants from Europe; and though religion and patriotism both plead for it, yet there might be found some minds weak or wicked enough, to prefer the murder of the little life left in our lands, to a slow process of resuscitation.

Forbear, oh forbear matricide, not for futurity, not for God's sake, but for your own sake. The labour necessary to kill the remnant of life in your lands, will suffice to revive them. Employed to kill, it produces want and misery to yourself. Employed to revive, it gives you plenty and happiness."

Here is a man who loved the land and his country. That these kind of men farmed, and wrote about it, should prove that every culture has a remnant that adheres to long-term thinking and looking for the truth. It may not surprise you that Taylor also sided with the large minority contingent who did not want this new country to have a professional military. They said it promoted wars and foreign entanglements. Their preference was a highly trained militia, but their concerns did not carry the day.

Benjamin Franklin, describing the difference between America and Europe, wrote in 1784:

"Great establishments of manufacture require great numbers of poor to do the work for small wages; these poor are to be found in Europe, but will not be found in America, till inlands are all taken up and cultivated, and the excess of people who cannot get land, want employment. . .

Industry and constant employment are great preservatives of the morals and virtue of a nation. Hence bad examples to youth are more rare in America, which must be a comfortable consideration to parents. To this may be truly added, that serious religion, under its various denominations, is not only tolerated, but respected and praised. Atheism is unknown there; Infidelity rare and secret; so that persons may live to a great age in that country without having their piety shocked by meeting with either an Atheist or an Infidel. And the Divine Being seems to have manifested his approbation of the mutual forbearance and kindness with which the different sects treat each other, by the remarkable prosperity with which he has been pleased to favor the whole country."

I just had to let the quotation go on to the religious part. In our time of political correctness, reading what the founders of this nation wrote, seeing firsthand what they believed, would certainly shock 99 percent of the school children in our country. We've come a long way, baby. Downhill.

In our love affair with everything glitzy and techno-awesome, parents have allowed themselves to be taken in by the prevailing notion that farming is a demeaning vocation. If we're going to make a great contribution to the world, we'd better do it in a laboratory or in Silicon Valley. May I be so bold as to suggest that people have been eating far longer than they've been punching computers? Is it possible that our basic need to eat nutritious food could be more fundamental to our life than the latest computer chip?

So my first reaction to the question about how to have kids who love the farm, but yet stimulate them with everything techno-glitzy, is to ask: "Why does it take that other stuff to make you proud of your child?" A soul search will usually reveal that Mom and Dad have this need to be accepted by their neighbors, and their neighbors don't regard farming as a vocation worthy of really bright kids. And so the peer dependency developed early in life carries Mom and Dad into a subconscious push toward stimulating the children in every direction except farming.

The stark reality is that these subconscious thoughts determine the way we interact with our kids. They create the subtle nuances of language, and determine what gets a big smile or a mediocre smile. I've watched countless farming parents, many of whom actually came to it late in life or from an unconventional approach, completely lose their kids because they overstimulated them in every direction except farming.

Every stimulant to which we expose our kids vies for their taste and attention. Every experience creates a desire or abhorrence. Proverbs 22:6 says: "Train up a child in the way he should go: and when he is old, he will not depart from it."

I've heard this rendered "Touch the palate of a child, and when he is old, he will not depart." The palate is the part in your throat that makes you swallow. When the doctor puts the wooden stick in your mouth and says: "Say Ahhhhh," and you quit saying "Ahhhhhh" and begin retching--he's hit your palate.

Life is full of stimuli touching your children's palate. In the Hebrew culture, before Gerber's, mothers would chew up food for babies. Then they would take that partly chewed food, put it on their finger, and touch it to a baby's palate. The baby would swallow and the sensation of course was satisfying. It didn't take long for a mother to just show a baby that index finger with food on it and the baby's mouth would fly open. That's the idea of this proverb. When we create a hunger and desire for what is right, the child will never depart.

The problem is that creating that hunger entails far more than reluctant obedience. It is more than 10 commandments. It goes right to the bottom of the soul, right to what turns you on and turns you off. When our kids get "religion" a couple of hours per week, but feed on pagan stimuli for 50 hours per week, we're not creating an appetite for the right things. We must control what goes in. And the fact is that everything can't go in.

That brings me to my second notion about the question we're examining. You can't do it all. Nobody can do it all. You can't be an expert on every subject. You cannot be an expert heart surgeon, expert dirt shoveler and expert steamship captain. Each area of expertise simply requires too much specific knowledge and experience. Proficiency in only one field is a noble aim, let alone numerous fields.

Perhaps a way to look at this is to realize that we can't give to every worthy cause. If any of us gave to every cause meriting donations, we'd all be broke. We can't go to every noble social function. We can't give every person what he might need to be successful. The point is that we must prioritize, and there's nothing wrong about prioritizing.

Just because I give to the water relief effort in Kenya instead of the Romanian orphan's fund does not mean I hate children. Just because I give to the neighbor whose house burned down instead of the virgin redwood lovers doesn't mean I hate trees. Our society's cause leaders have done a masterful job at making us all feel guilty. I'm supposed to feel guilty for the dirty rivers. I'm responsible for racial discrimination,

Vietnam survivor's syndrome and an unwed mother's predicament.

The daily media barrage has made us all far more acutely aware of every malady in the world, and heaped upon us a burden I don't think humans were meant to carry. When all of our senses are assaulted with a hundred griefs per day, our psyches just break down.

We've had this inordinate burden to heal all the wounds, solve all the problems heaped upon us, and its ancillary notion that we must involve ourselves in everything. So we live these hurried, harried lives, going from one function to the other, and never putting down roots. If that really solved problems, we'd see divorce, dysfunction and drugs dropping in our culture. But for all our activity, they are going up.

Clearly, "busyness" doesn't establish roots. Flitting from this activity to that one creates children who will be just the same. They won't develop an appetite for home and hearth because Mom and Dad taught them, subliminally to be sure, that to really live you need to go somewhere. Who wants to be stuck at home? I do.

Is that too radical? What, pray tell, is wrong with Mom and Dad raising children who want to carry on the family farm business? Who want to continue serving in their fellowship, dealing in commerce with local merchants, and carrying on a tradition of moral purity, neighborly love and stellar business ethics? Why is that not as noble as inventing a new jet fuel or being a senator? I'm not suggesting that these other two possibilities are not noble. I'm only suggesting that the multi-generational family friendly farm is at least as noble as their more glitzy counterparts.

And when we ask questions like "How do we get the kids interested in the farm but make sure they partake of all the glitzy things society has to offer?" we belie our philosophical foundation that assumes the glitzy stuff is necessary or germane to success. Or even that it's desirable. Our farm kids can easily have all the experiences and stimuli they need to be successful farm kids. Why does a ballet and academic prowess rate higher than a happy, productive, community-influencing family farm?

The answer is you can't do it all, and you needn't even try. And if you do try, you'll never get the kids interested in the farm because the lure of riches and fame are too strong. Oh, they'll develop an appetite for something, alright. But it won't have anything to do with the family farm.

Lest I be grossly misunderstood, I'm not suggesting that we hide away from society like some agrarian monastic order. And the chapter on hermits certainly proved that.

But I am suggesting that we carefully regulate the stimuli so that we do not compromise our message. For example, if our children associate going to a museum as more valuable than producing nutritious food, we're in trouble. We must understand that regulated exposure is healthy. Exposure, yes, but metered exposure.

In my opinion, an adolescent child expert at building fence, shoveling dirt, dressing animals, running a livestock business, building compost, managing cattle, loading hogs, identifying diseased trees, and driving farm machinery will inherently have less computer prowess than an adolescent who spent five hours a day playing with computers. Am I opposed to computers? No. All I'm saying is that there is not enough time in the day to become an expert at everything worth becoming an expert about.

I'd love to know more about chemistry and physics. I'd love to know more about how this computer works on which I'm writing right now. I'd like to know the members of the local orchestra too. But I can't do it all. Neither can you, and neither can your kids. To be successful, I must stay riveted on target and reject things that would sidetrack me. Just because you haven't visited the local art show or learned what every historical marker says does not mean you despise beauty or disdain history. We all must pick and choose what will get first dibs on our time.

And by all means we need to get over our own dependency on others' approval. It will take a radical change in world view, a reassessment of what's really important. But if we never arrived at a deep contentment, a complete satisfaction with the agrarian life, we can't expect our kids to adopt it out of the blue. They sense where our heart is, and if our heart is somewhere else, that's where they'll head. We can't play a charade all day every day. The kids know. They sense where we are. And if we're not locked onto family friendly farming, they'll be looking to where our laser beam is pointed.

Being satisfied with where we are is unbelievably liberating. I see parents chasing here, chasing there, trying to make sure their children have seen the latest greatest everything. We have a friend who calls this "street smart." Teresa and I have no desire to have "street smart" kids. If they have no idea what the latest icon on designer wear is, we couldn't care less.

Our own kids have never had a Happy Meal. They have not a clue what's being offered at the burger joint to get kids to come. They're not well versed on Hollywood. They think men's earrings and girls' tatoos are disgusting. When Daniel went to a fishing camp once, he was teased unmercifully for tucking in his shirt and wearing his hat with the bill heading forward. They have our values and our tastes. After all, they are our children.

Does this make our kids unfunctional in social situations? Remember, Daniel was president of Virginia 4-H. Rachel is president of her Ladybug Club. They don't dress funny or talk funny. But they do have a steadiness about them that is magnetic to other young people. And steadiness comes from roots; roots come from staying in one place a long time. Period. And roots come from unwavering devotion to a cause. Our farm ministry, to really put down multi-generational roots, requires too much time and attention to uproot it with glitzy entertainment and shallow cultural values.

We've tried to stimulate the kids with experiences that reinforce our notion of what is important in life. When we travel, we visit successful farmers. We visit museums and living history villages. We like old crafts and appreciate heritage arts. But even these we regulate.

Otherwise, our kids might want to go and be costumed interpreters there. It's not wrong to be a costumed interpreter at a museum. Please don't misunderstand. That's just not where we are as a family friendly farm. We are driven about this model, and we believe it can be done.

I grew up without a TV. Imagine where that put me with the culture of the '60s and '70s. But I had roots. Right here. And we covet that for our children and grandchildren. They won't all stay, but we're going to do our best to make sure they all love it here and have the opportunity to be here. God helping us, we're going to stay with it. And we think that's noble.

Balancing Stimuli: You Can't Do Everything

Chapter 28

Family Council

How do we keep the farm business family friendly? The only way I know is to have frequent family councils. Set aside a couple of hours at a time to hash things out. Sometimes everyone is going different directions and it feels like it's spinning out of control. That is when you need to sit back and just communicate.

This is different than just fleeting informal discussions during the day. This is a genuine, formal planning time. Virtually every morning our family sits down and creates a "To Do" list for the day. This helps everyone stay on track. The jobs are divvied up according to ability or desire and everyone starts in on the assignments. But the family council goes beyond this because in the day-to-day hustle and bustle of activity, little things can add up that need more air time. Not necessarily resentments or disagreements, just hunches, feelings about where we're going.

Of course we have our days when one of us says: "Let's just shut it down. These apprentices are driving me crazy. We don't need to produce what we're producing in order to make a living." But then when we come together--just family--and discuss things, we always come to a consensus and part on the same page. It really is amazing, and as the children get older, it becomes more and more crucial. Because now they reason with adult thinking, and their insights are not only valuable, but often more perceptive than mine when it comes to the nitty gritty details of chores.

One element of these family council sessions is to discuss additions or subtractions to the business. Often one person wants to push ahead on a project while another wants to retreat on it. If we can't come to a consensus, we just put the whole thing on hold until next time.

It's easy to overrun your headlights in the business, especially when Dad is the entrepreneur. Take it from one who continues to learn this lesson--you can't run away from your family. I don't care how much you believe in leadership and chain of command, you cannot run too far ahead of your family or they will be lost. I've seen this especially in middle-aged men who shut down their urban lifestyle and move to the country.

I don't know how many pensive wives have called us and said: "My husband is a wonderful man. He's always been level headed and a hard worker. But he's got this idea of selling out, quitting his job, and moving to the country to farm. Neither of us has ever farmed. I'm really worried. We have two daughters. Can we really make a living?" It always bothers me when the wife makes this call.

The same Bible that tells wives to submit to their husbands tells husbands to live with their wives according to knowledge. At the same time, the balance to "children obey your parents" is "fathers, provoke not your children to wrath." Unfortunately, too many men emphasize the one side without appreciating the other. Sometimes the wife is the one pushing for a change of life. She's tired of not being able to build a nest for the family and wants a change in lifestyle. But the husband is scared to death that he won't be able to find employment.

A "fullness of time" exists for all these decisions. No husband, wife, son or daughter can maintain a family friendly atmosphere while running too far ahead or lagging too far behind the rest. This family council I'm talking about is not a time for preaching. It is not dominated by any one person. It is a time to listen and to understand. Resolution and consensus are the goals. Here are some red flags that signal it's time for a family council:

- "I hate this farm."
- "You never listen to me."
- "We only do what *you* want."
- "You don't love me."
- "I hate this ____________ (fill in the blank)."
- "Why can't we get organized?"
- "Why do we have to do _______ (fill in the blank) all the time?"
- "I never get to do what I want to do."
- Emotional outbursts, indicating deep trauma over a situation.

I don't care who says or does any of this to whom--whether any of this is proper or religiously acceptable is not the point. The point is that these kinds of statements reflect deep hurt and disappointment and had better be taken seriously. And when your child says: "I never get to do what I want to do" it doesn't help to respond: "That's a lie and you know it. Shut up."

A better response is: "What would you like to do?" This pulls the pressure off and opens up the discussion. And I don't care how angelic or demonic your children are, their feelings are real and they need to know you will give them an audience. Emotions are powerful tools for both healing and conviction; perceptions are real. We must give the other person a fair shake for how he arrived at his feelings in order to keep lines of communication open.

On a personal note, my perfect daughter, Rachel, began saying disparaging things about the appearance of the yard scape. I'll admit I was a little slow on the uptake, and my first reaction was to chide her about not being content. But then Teresa, as is often necessary, helped me understand that to our little artist, my hodgepodge of portable outbuildings around the yard perimeter was a real embarrassment.

So we held a family council. We took pencil and paper and let Rachel rearrange everything. Don't just listen to complaints from the children; turn their complaints into action. Complaints without positive solutions are out of order. Except when your wife is just unloading. Teresa has learned to preface her unloading times with: "Now you don't have to fix anything. I'm not asking for solutions. I'm just going to unload for a minute." It's been a hard lesson to learn, but I'm finally getting the message.

The session was incredible, and Rachel solved all sorts of

problems. Now she owns the landscape. And that is what the family council is all about--ownership. When everyone signs off on the projects and the business direction, and everyone has had input into the decisions, everyone shares ownership. It's no longer Mom's deal or son's deal. It's "my" deal.

Those of us blessed with an entrepreneurial bent are always ready to start something new, often not realizing the effort required to maintain what's already going on. The family council helps to keep a rein on all of this, but also provides an opportunity for the one wanting to do something new to express how it will augment the business.

Another source of contention within the family farm is the allocation of money and labor to projects considered domestic, or household, and the ones considered germane to the business. This tug of war occurs all the time. The farm needs a new tractor and the house needs a new washer/dryer. Which one gets satisfied quicker? If the farm constantly gets the latest and greatest while the home gets leftovers, pretty soon dissatisfaction will set in, then resentment, then disunity.

The old saying: "When mama ain't happy, ain't nobody happy" describes this relationship well. I have all sorts of excuses to put Teresa's household needs on the back burner: "They don't augment the business, but a tractor will." It sounds reasonable to me, but what she hears is: "My projects get first dibs. Your needs aren't important." Balancing the outside/inside projects is essential to maintaining unified passion about the farm business.

Here again, the family council comes into play. When we make our list of big projects for the coming year, everyone has input. We must avoid compartmentalization. One of the things I hate to hear is "his project" or "her project" instead of "our project." These are little signpost pronouns we need to be listening for that help us know whether the family members have ownership or feel stepped on.

Family council can be as formal or informal as you want to make it, but it must be professional and it must be private. It must not include employees or non-owning relatives. The privacy requirement insures complete freedom to air differences. Otherwise, we'll hold back because we don't want to air dirty laundry for the world to hear. Sometimes getting this private time can be difficult, but it is essential in order for the meeting to function properly.

It can be scheduled far in advance or be fairly spontaneous. Ours tend to be spontaneous as soon as we see we'll be alone for a couple of hours. In other words, just taking advantage of a normal scheduling situation. But don't be afraid to actually have a monthly time if that's the only way you can get it done.

Why isn't the family council more common? I don't know for sure, but I'll guess for the same reason we tend to not pay attention to it until an acute flare-up signals its necessity: laziness and fear of hearing others' assessment of us. When everything is rocking along fine, the easiest thing in the world is to let it run on autopilot. But we've learned that there really is no such thing as autopilot in family business because people, unlike clouds, wind, propellers and wings, have memories and emotions.

We don't just respond like a baseball to the force in a pitchers' hand. Instead, we're analyzing how hard we're gripped, how fast is the acceleration, and why we're thrown that way instead of this way. Nothing involving people can be put on autopilot. This is why all the functional thought models have a monitoring element. Reassessment is the key to maintaining momentum in a given direction.

The other reason people don't do family councils is because we fear honest appraisals about our thinking and actions. If I get off in my little world and lose touch with projects others are working on, I'll be called into account for being disinterested. I don't like my son or daughter, much less my wife, telling me I'm disinterested. And yet that is exactly the truth and an honest appraisal more often than I'd like to admit.

Accountability alone is a good reason to have the family council. When one family member is fiddling around or not holding up his end of the load, everyone else can agree and in the professional climate of the moment, it's not just finger pointing from one sibling toward another. An agenda really helps to make things appear professional.

In these meetings, no busywork should be allowed. Period. No cross stitching, no crocheting, no reading or earphones. Undivided attention must be focused on each other, solving the problems and moving down the agenda. Establishing this mutual respect clears the way for creative, sensible discussion rather than ranting and raving emotional outbursts. And by the way, it's okay to say to someone: "I'm angry with you." A shared responsibility exists in the offending arena.

That statement should not offend anyone. And if it does, the offended party needs to get a life. I'd much rather have someone tell me an honest statement to my face, than backbite and gossip when I'm not in the room. Or worse yet, develop judgments and perceptions about me without giving me an opportunity to defend myself. People who are easily offended will never experience deep accountability, nor the satisfaction of overcoming weaknesses. People around them will always walk on eggshells, and their friendships will always be shallow.

Plenty of books have been written about techniques and dynamics in group discussions, but here are a few tidbits you might find interesting:

- Give each participant slips of paper worth one minute each. These are given up during discussion to keep one person from dominating the dialogue.

- Be cognizant of temperament types: sanguine, choleric, melancholic and phlegmatic. Each will hear and articulate things differently.

- Each succeeding speaker must summarize the previous speaker's comments before starting in on his own. This facilitates listening.

- Use flip charts. Writing down comments for all to see empowers participants.

- Brainstorm often. The operative word here is "nonjudgmental." Brainstorming keeps out all qualitative comments until all the ideas are on the table.

- Proceed by consensus rather than simple majority. This protects the minority and creates a real commitment to push through to real solutions. You want to see all heads nodding before calling the problem "solved."

While I believe the family council becomes more and more important as the children get older, I realize now that even as toddlers our children were involved in this because we did not "protect" our little ones from conversations many adults would deem over their heads. Children pick up a lot more than you might imagine, and I'm convinced that the common sense exhibited by our children today is a direct result of watching Teresa and me wrestle with problems in the early days. How we talked to each other, how our thought processes worked: these procedures have been inculcated into the fabric of these young people.

By the way, I don't use the term "teenager." That term was not even in the dictionary until the 1960s, and represents an assumed generation gap. I think the term demeans young people and gives this age group license to act foolish. When we adults use terms that assume adolescent foolishness, that's exactly what we'll get.

Family council is the glue, the measuring stick, the laser beam---use whatever metaphor you want. But it is an essential technique to keeping everyone on track toward a commonly owned goal. It is critical for properly functioning, multi-generational, family friendly farms.

Family Council

Chapter 29

10 Deadly Destructive Deeds

At the risk of being redundant, I want to offer my list of ten deadly destructive deeds. This is a book about opportunity, empowerment and happy families. I've purposely tried to give it a positive flavor. But often we learn more from our failures than our successes. Negative examples are powerful teaching tools.

In that vein, indulge me one chapter to capsulize what I consider the essentials to destroy a family friendly farm specifically, and a functional family generally. These are in order from least devastating to most devastating.

10. **CONVENTIONALISM**. Parents who go with the flow produce children with neither anchor nor rudder. Religion, education, diet, recreation, vocation--no area of life should come outside the scrutiny of the conventional paradigm. Just doing what everyone else does is a sure way to keep our children from doing anything valuable with their lives. They can just punch time clocks, tell the same jokes and fill up their 24-hour days with the same nonsense as the average person.

People who make a difference don't do what everyone else does. Statistics tell us that 75 percent of high school graduates never read another book the rest of their life. A group of six attending any local political party caucus can take it over. Conventionalism handicaps our children for life.

9. **SECRECY**. "We'll talk about that later" and "Mommy and Daddy have to have a conference" are telltale signs that this is practiced in the home. Children learn quickly that parents can't trust them with their thoughts, so children do not trust their parents with thoughts and become scheming, conniving individuals. When our discussions are shrouded in secrecy, our children learn to disrespect us because we can't speak forthrightly.

While the rationale for this may be a unified front on Mom and Dad's part--and that is incredibly important--the children must not get the idea that their parents are always talking about them. Thinking about them, yes. But talking about them with dubious intentions is out of line. This is different than typical pillow talk, where Mom and Dad deal with the important issues facing sibling interaction.

Children need to know that Mom and Dad are straight up with them. And I would suggest even that seeing Mom and Dad disagree a little but quickly come to consensus can be a powerful learning experience for the children. Watching how Mom and Dad disagree like adults is a valuable model--as long as Mom and Dad come to consensus.

8. **MANHANDLING**. Few things irritate me more than to see parents physically bossing their children. Whether it's grabbing cheeks, gabbing a shoulder and spinning a child around, or jerking on an arm. A child has a wonderfully padded spot for absorbing

physical punishment. And lest anyone think I am archaic for endorsing paddling, I also endorse whipping posts instead of jail.

Nothing is more dehumanizing and dysfunctional than prisons. My conservative friends strut their "tough on crime" message, building more jails as if that will solve anything. The philosophical foundation of prison is flawed. Quick punishment and the second chance work, not penning people away from society in incubators for dysfunction.

I cringe when I see parents slap faces and jerk little tots around. This simply offers approval for bullying. If a child needs a spanking, then administer it. Otherwise, keep your hands off.

7. **DISINTEREST**. Parental disinterest in the children is more widespread today for two reasons. First, parents aren't around the children as much because work is outside the home. Second, children have more distractions than ever before with electronic game entertainment, television and the Internet.

As a result, it's very easy for parents and children to live in two different worlds that never intersect. The kids have their life, Mom has hers and Dad has his. The non-custodial parent in a divorce has an especially big disinterest problem to overcome. "He just walked out" is a sad commentary on our culture. It is despicable. Oh, people are interested, alright. Just not about the right things.

6. **DOTING**. Doting parents have no idea the harm they do to their children. They think it's love, but it is teaching a selfish, egocentric world view. Showering the children with toys, supermarket treats and the latest designer clothes certainly does not produce selfless, disciplined, productive citizens.

We are certainly entering an age when this is more common. The pendulum is swinging away from the "children are to be seen and not heard" to "children call all the shots." We worship children today, as if they've become kings. This is the unfortunate reaction to child sweatshops common during the early days of the industrial revolution. It's another guilt abatement program, and it yields bitter fruit.

Any post one-year-old who sasses his parents will be a handful when he hits sixteen. A smart-mouthed toddler is obscene. Temper tantrums are not cute; they are the practice field for sixteen-year-old disrespect.

5. **PERFECTIONISM**. Never being able to do things well enough destroys the self-worth of children. Generally springing from a parent's poor self-image, or feeling of inadequacy, perfectionism destroys team spirit and manifests itself in older children who only help out when coerced.

A sure sign of perfectionist families is children who are reluctant to help when a job needs doing. Eager-beaver kids don't come from perfectionist households.

Little fingers have poorly developed motor skills for making perfect lines and letters. Little eyes aren't as sharp for close work, like spotting that little spot of egg on the dinner dishes. Oh, parents, please don't rip the zest out of your children by demanding perfectionism.

4. **AUTHORITARIANISM**. "Because I said so" should never be uttered in a home. It is a copout and creates a sense of dominance just for the sake of dominance. Authority must be balanced by love. While parents should demand obedience from the

children, they also need to foster a spirit of listening.

Even God allowed Moses and others to change His mind. Children who believe they have influence with parents will grow up sensing that their contributions are taken seriously and have real value. When their viewpoint is dismissed as nonessential or irrelevant, as is often the case, they finally keep their thoughts to themselves and develop a resentment toward authority.

This arrogant domination mentality also bears fruit when children hit adolescence and become sassy and disrespectful. Control is the operative word here. Parents who grew up in home practicing conditional love, or in homes shy on praise, often feel the need to control their children to make up for a subconscious feeling of inadequacy. Weak adults often over-control their kids with authoritarian dominance.

3. **BAGGING.** "You always act that way!" "You never treat your sister nicely." These attack the personhood rather than the deed, and bag the child. Attacking the individual creates a hopelessness because while the deeds can be punished or change, the individual is more difficult.

This again goes to the self-image problem so many young people have. It's as inappropriate to dote and give unmerited rewards as it is to bag the individual. No matter how heinous the crime, preserving dignity and maintaining respect for the perpetrator must be the goal. Another strike against jails.

2. **SCREAMING**. This represents a lack of self-control on the part of parents, and certainly does not engender respect. I heard an excellent metaphor that makes the point. Imagine if you were stopped for speeding and the policeman, instead of quietly and

professionally writing you out a summons and wishing you a good day, jumped up and down, screaming and waving his arms at you. "You stop that speeding right now! Do you know how many people you endanger driving like that? What's the matter, are you blind or just ignorant? You don't pay attention. Get with it!"

We could have fun with this little ditty, but you get the picture. People would stop just to watch the show. After awhile, you'd begin to laugh. If that were the normative procedure, how many people would seriously care about their speeding? Nobody.

The only thing worse than toddler temper tantrums is adult temper tantrums. Instructions, punishment, commands--all these should be meted out under control. If you're losing it with your kids, stop and count to ten. Never punish when you're out of control. I don't think it's wrong to tell your little child that you are angry with him. That is the truth. But tell him you are angry and disappointed without screaming.

1. **INCONSISTENCY**. Here we are at the big number one. Inconsistency is the worst violation in parenting, and all of us who have ever had children are guilty. What is wrong today should be wrong tomorrow. The limits today should be the limits tomorrow.

Mom should agree with Dad. No child should be able to play one parent against another. And the worst is teaching our children that we don't mean what we say by using the threat and counting message: "You'd better be in bed before I count to three . . ." "I've told you three times to stop that. Now stop it."

My heart aches when I see parents abusing their children this way. All it shows is that the parents have no idea about how to raise children. They have no idea that the toddlers are trying to control the situation, and too often the little guys win. Children want to know where their boundaries are. They want to know you mean what you say. They want clearly defined, safety-engendering fences that offer more playground freedom.

I read once about a study done where kindergartners were put in a playground without any fences. The children just huddled next to the classroom building, afraid to go out and use the equipment. Then they built a fence around the yard and had another group of children exit the classroom door. These children immediately spread out in the playground. The fence offered security and clearly defined borders. This is what we as parents need to give our children. They yearn for it, and we must be willing to build those fences for them.

There they are, the big deadly ten. I'm sure that as you read this list, some of these items hit home and others you dismissed as inapplicable to you. Aren't you always amazed at the things people overlook? The office deadbeat cannot imagine he's lazy. Everyone else knows it but he has no idea. Or the church gossip. Why, he wouldn't think of calling himself a gossip. Everybody knows it, but he's completely blind to it.

So how do I know how I fit into this list? That's where accountability between spouses and friends comes in. Community and its proper functioning. The individualistic, democratic culture we've created keeps any of us from counseling with others to get honest appraisals of our weaknesses. I encourage you to go to your friends and ask them how you measure up on this list. If you have children older than about eight, ask them how you measure up.

If we don't seek, we won't find. This is one reason why the master plan has a man and woman raise children. It takes both.

Each offers insights, and carries a makeup that complements and offers balance in the home. Each of us has violated these dastardly elements, and the kids are resilient. Fortunately, they can handle occasional slips. But they can't handle habitual offenders. Let's all dedicate ourselves afresh to watching out for these deadly destructive deeds.

Chapter 30

Nutrition and Lifestyle

I'm sure some parents with disabled children, whether physically or emotionally, may be thinking at this point that I am awfully glib and unappreciative about human limitations.

"Doesn't he understand my child has Attention Deficit Disorder and just can't follow instructions?"

"Doesn't he understand that my autistic child can't be left alone to weed the peas?"

"Doesn't he understand that my Down's Syndrome child will never operate his own business?"

"My 12-year-old has leukemia. What does that do for our family business?"

Human tragedy often is simply inexplicable. I certainly don't have the answers to all these situations. But let me offer a few thoughts so that at least you will know that I have role played some of this.

At no time during the 10 commandments listed earlier in this book have I said: "Give your kids good nutrition." Perhaps that should be one, but it would be a little out of sync with the rest, so I did not put it in. But we cannot underestimate the importance of good nutrition. This is not a human nutrition book, but this is certainly as important as our spiritual food.

When I was college age and planning to return to the farm to produce natural food, the fundamental and evangelical Christian community absolutely excoriated the natural foods movement, labeling it a cult. This was in the mid 1970s, just on the tail end of the hippie back-to-the-land movement. The counter culture movement, viewed with contempt by the entire conservative community, brought a sense of environmental stewardship and reintroduced a sense of awe and reverence toward the living world.

The religious right was busy feeding Twinkies to their kids and then praying for their healing when they got sick. Conservative families denied any link between physical and spiritual. As they pored over systematic theology books they consumed soda and Spam. And they castigated liberals as being out of touch and inconsistent.

Meanwhile the liberals were casting off all the ethics and morality that characterized Judeo-Christian success, but embraced a respect for the creation. Paul Ehrlich prophesied the end of civilization. His prophesies were so off base it's amazing anyone still listens to him. The religious right worshiped doctrine to the exclusion of the physical while the religious left worshiped creation to the exclusion of doctrine. The "do your own thing" mentality, encouraging an anchorless, boundaryless existence, was as wrong as the conservative straitjacket and utter disregard toward creation.

This began for me the ideological homelessness that has characterized me ever since. If I go speak at an environmental conference, I can tell that the audience assumes I probably vote for the Green Party, endorse abortion, hate capitalism and despise corporations. People love to pigeonhole other people. That is why I tell them I'm a Christian libertarian capitalist environmentalist. Each one of these quarters brings to the table some good elements to temper the negative aspects of the other three. We tend to jump to simple solutions and too often things are not that way.

I was in Canada attending a banquet at an organic farming conference when one of the fellows seated at my table said: "I hate Christians." I can't remember what led up to the comment, but it came at me like a 2x4 and sounded like a train of thought I probably ought to pursue. After all, I are one.

I asked him to explain what he meant. As it turned out, he and a cohort had just returned from a six-month videography tour in Africa. I don't remember what they were video taping, but his point was that they visited impoverished village after impoverished village, often right after a shipment of western aid had arrived.

Generally, these were container loads of clothes, cosmetics and nonperishable foods sent by American Christian missions organizations. Although these villages were impoverished, the political climate was a mess and living conditions generally were harsh. There were always go-getters and entrepreneurs who created a local economy. The clothes these local vendors sold were not western designer stuff, but they were functional nonetheless and supported a vendor or two and their families. These vendors had their booths and represented a semblance of local commerce and industry. They were the best these villages had.

Along came this container load of goodies from the west. The villagers flocked to the freebies and quit patronizing the local vendors. The local vendors went bankrupt because, at best, they operated on a shoestring, and once the goodies were gone the local economy was worse off than had the freebies never arrived. What was worse, since these vendors represented the most aggressive villagers, they would often turn to crime after losing their business.

And, in fact, these folks who had just returned from the area, said that many of the pillagers in the countryside were these displaced businessmen. The ones who had tried to prop up things the best they could now become the ones terrorizing the region. I was absolutely amazed when he finished his account. I thought of all the pleas for help sitting on my desk at home, and all the grief I felt for these poor people. But then I realized just how arrogant we are when we swagger into situations and throw around our: "Well, you just need to do this and this. If you'd just think like this instead of that, blah, blah, blah." How shallow.

And so I appreciate how shallow it sounds when, with two incredibly healthy, well-adjusted children, I sit here and speak to families whose pain and struggle is beyond anything I can imagine. My friend Kyle has an incredibly wise saying: "Once you hear about someone else's problems, you'll always take your own back." I know that when I've faced government bureaucrats confiscating our beef, accusing us of selling uninspected meat and threatening to close our farm, I've gone through some deep water. The fact is that all of us have gone through our valleys.

And once we start comparing valleys, if we're honest, we'll always take our own back. When we're in the midst of the valley, we always think ours is deeper and more foreboding than anyone else's, but it's really not. People who think I'm a successful writer and farmer don't know the daily struggle with people wanting information, trying to maintain privacy and serve people at the same time. Oh, it sounds romantic on the surface, but it has its valleys when the kids finally hit me with a 2x4 and say: "You're helping all these other people, what about us?"

Time for a reality check. I feel miserable, like a failure in the things that matter most. I have a temper problem that puts me into the needing forgiveness category too often. Okay, let's all get radically honest here. Forget the charades. Forget the smokescreen. Let's all just admit that we have problems, valleys and seemingly insurmountable obstacles. And let's get on with life. Read about Helen Keller. Read about anyone great, and you'll read about valleys most people never enter.

King David wasn't born royalty. He didn't grow up in Jerusalem. And when he made his debut, he faced a man twice his size who graduated first in his class from the military academy. His foe was a combination Sumo wrestler and Sherman tank. But David flung the rock and then spent the next several years running from a jealous King Saul. Then, when he finally reached the top, he committed adultery with Bathsheba. Worst of all, he got her pregnant. So he had to murder her husband and then the child died. What a horrendous valley for this man who knew better. But he had a soft heart, and that made all the difference.

He also set about to make things as right as he could. And I think that's where I'd like to touch those of you with these seemingly insurmountable obstacles. First of all, forgive. Let your heart soften toward the event or person that is keeping you hardened and all bound up.

Then let's look at alternatives. I don't have solutions to

everything, but I know that far more real solutions exist than most of us realize. And this brings us back to the beginning of this discussion--nutrition. The link between many emotional, behavioral and physical problems and good nutrition is growing more profound everyday. Wise Traditions magazine, published by The Weston A. Price Foundation, had an article by a medical doctor showing the direct link between autism and vaccinations. But it doesn't stop there, and offers vitamin A and D in cod liver oil (not synthetic) and other natural foods as a significant solution.

The alternative community abounds with links between water fluoridation, synthetic preservatives and food colorings and a host of toxins we ingest when eating out of the conventional supermarket. Not to mention the epidemic in food-borne bacteria. The solution to that problem is not in irradiation, but in pasturing animals and getting them out of fecal factory farm concentration camps, which create the pathogen-enhancing environments in the first place. If your family lives on carbonated soft drinks and Spam, you'll have children prone to ear aches, runny noses, allergies and skin problems. Look, something as simple as breast feeding has been shown to eliminate many allergies.

I don't have time to go into all the nutrition-related illnesses here, but believe me when I say that a whole world exists that approaches life from a nonmechanical, nonindustrial world view and that offers real solutions. I hope no family reading this book will eat white bread for sandwiches and Lucky Charms for breakfast. But if that is your nutritive base, all sorts of emotional and physical maladies will show up.

We had a customer once who cooked for the local school system's behaviorally challenged program. These were kids so out of control they could not be mainstreamed. It was an institutional setup. He found that if he withheld refined sugar from them, 90 percent straightened up within a couple of days. The problem was they went home on weekends, loaded back up on sugar, and came back as bad as before.

Back when I was a newspaper reporter, I attended the dedication of an athletic field at one of these state operated adolescent institutions. A couple of the youngsters began getting out of control so a staff member turned around and asked if they needed some "behavioral reinforcement?" I waited to see if he'd take them into the back room and apply the board of education to the seat of knowledge. No way. He pulled out a two-gallon bucket of hard candy, and gave it to these kids, who consumed it by the handful.

Anyway, my friend the cook finally went to the county education board and told them they didn't need this million-dollar program; it was strictly a home dietary problem. The board, of course, was getting a lot of state and federal money to run the program, and it required an administrator at a central office to oversee things, employ a couple of psychologists and counselors to keep the division's employment nice and fat--you get the picture. No, they weren't interested in terminating the program.

This is why I have a real problem with the liberals who want salvation by legislation. They want to throw more money at education, and call people like me "anti-education" when I don't think we need a tax increase to pay for more conventional educational stuff. Both conservatives and liberals love government agriculture programs that fund their pet projects. Farmers who vote at both ends of the spectrum enjoy receiving their subsidies from the public trough.

I hope my assessment of these problems has not sounded

flippant or uncompassionate. But when confronting evil and willful ignorance, I confess it's difficult to restrain myself. In my opinion, real child abuse is not seeing children working in far eastern sweatshops, slowly bettering their station in life. It is children eating supermarket-bottom feeder-pseudofood. That is real abuse. And to get all bent out of shape about the one without confronting the other simply shows just how small minded and myopic we are.

Some incredible work is being done by Jo Robinson, author of Why Grassfed is Best. She is showing how the additional bends in the fatty acids of pasture-raised meat, poultry and dairy products gives additional efficiency in brain function. The nutritional component is not incredibly difficult to fix--and another reason why the farm is inherently more capable of producing healthy kids. By the way, the bends in these fatty acids are impossible to duplicate synthetically, although the natural foods industry is marketing pills purporting to be the same thing. They aren't honest either.

Beyond nutrition, staying at home greatly minimizes the risk of juvenile illnesses. Exposing children to the assault of foreign pathogens could probably also be labeled child abuse. Documentation abounds linking child sicknesses to immune system overload in classrooms and other people-dense places. Especially in the early years, we must minimize exposure to strange bugs. When I say home is a haven, I really mean it. This is not some ludicrous, pie-in-the-sky idea.

People who think culture cannot be influenced from home are missing the point. We're talking here about very small children. We're talking about diet. The farm business functions in the community and in commerce. We'll talk a lot more about that later. But right now we're dealing with minimizing the risk of unnecessarily handicapping the family or any of its members. I do not believe that a child suffering harm due to pathogen overexposure is something God wills on that child.

Again, I don't have all the answers. But to say that if I neglect my child and she drowns in the swimming pool it must be God's will is to deny my guardianship any responsibility at all. Why not just let the kids run wild and let God protect them? Now I've probably raised the ire of all the hyper-sovereignty folks, but we're not going to solve the Calvinist-Armenian debate here. I can't reconcile it, so I'm willing to admit that it's beyond me. Why do we have to solve every debate?

The point here is for us as parents to act responsibly and to realize that this family friendly idea encompasses a whole lot more than our child learning to say "Yes sir" and "No sir." If we're in charge, then we're responsible for where we send our children and what we put in their mouths, who we go to for counsel and how we view the world. It all goes together in producing a well-adjusted young person with a passion for truth, a zest for life and a will to work.

Finally, the farm certainly offers as many or more opportunities for handicapped children to be an asset to a family business than any other vocation. Intellectually disadvantaged children can do many simple tasks. And that's one of the whole ideas behind the family friendly farm. When we begin structuring it with production, processing and marketing models, the systems we choose have a direct bearing on how many "workstations" non-adults can have.

The factory farm offers very few of these workstations. But one that requires more people, hands-on activity and creates a higher gross margin per item produced offers more work niches for special people. Landscape diversity, production diversity and patron diversity creates room for these special people. And that is family friendly.

I wish I could make all the hurts go away. I wish I could make my own hurts go away. But we can't do that. We must take a charitable look at our problems, realize they really are no worse than anyone else's, and move toward objectives that will put us in the elite ranks of over-comers. You can't solve my problems and I can't solve yours. And we sure can't solve all the world's problems. But if you get the best side up on yours, that will create a ripple affect to your spouse, your children and your neighbors.

Few things bring tears of joy to my eyes faster than watching an enthusiastic parent-child team work together. How many do you know? I'll wager you know more that don't work together than do. When they work together, they're beautiful. But when they don't, they're ugly beyond description.

I sincerely hope the previous section dealing with the 10 commandments of making your kids love the farm will challenge us all afresh to rise to this most fundamental of human tasks. The responsibility for loving the children into our farm rests with parents, not children. Parents, are we up to the task? I pray so.

A Sacred Work

Chapter 31

Industry vs. Biology

Sometimes it behooves us all to step back and look at the foundations of our paradigm in order to give us greater conviction in its defense. The philosophical underpinnings of our views are often easier to defend than specific details.

For example, I have debated agri-industrial darling Dennis Avery, author of Saving the Planet with Pesticides and Plastics three times in public forums and he is no dummy. A retired USDA big wheel economist, a Ph. D. and spokesman for everything genetically engineered, irradiated or confinement reared, he is articulate and likable.

Statistics, data and details flow off his tongue fast enough to paralyze the most intrepid debater. Certainly some of it seems twisted, illogical, or even contrived, but his appearances on the evening news with Peter Jennings, numerous national talk shows and constant speeches before prestigious scientific and agribusiness conventions attest that he is no pushover. He plays argumentation hardball.

But I have found a soft underbelly--philosophy. I look forward now to our next exchange. What is more important, honing this line of thought has actually made some diehard anti-biological folks turn their heads. All of us need ammunition in this war of ideas, this clash of paradigms between food production that stimulates earthworms and one that destroys earthworms. To be able to articulate our position well should be the goal of all who espouse the eco-friendly approach.

I am borrowing the industrial approach heavily from agriculture economist John Ikerd from the University of Missouri. He says the four pillars of industrial paradigms are:

1. Specialization
2. Simplification
3. Routinization
4. Mechanization

In contrast, nature's pillars are:

1. Diversified
2. Complex
3. Flexible
4. Biological

Let's take these one by one and deal with the ramifications in our world.

SPECIALIZED VS. DIVERSIFIED

The secret of the industrial paradigm is the interchangeable part. Furthermore, machinery and workers build up a skill level doing specialized tasks. Generalists are outside the industrial box.

The idea is that if each person does only what he is exceptionally good at, we will all become wealthy because there won't be any lost motion learning new skills.

A corollary principle is compartmentalization, which extends aggressively into our educational system. We have science, history, English and math, for example, and the subjects do not intertwine. That is why I can graduate with a degree in history without a clue what economics played in civilizational cycles, or knowing the megapolitical implications of the microscope.

In our university system we have Ph.D.'s who know more and more about less and less. Recently I attended an extension conference and listened to an agriculture economics specialist lecturing about whether or not to buy feeder calves or sell hay, assuming you had a surplus of hay. He had elaborate formulas for determining when it was better to feed the hay through cattle or just sell the hay, with supporting price differential charts and high-falootin computations.

At the end of the lecture, I asked him if he had thought about the fertilizer benefit from manure derived from feeding the hay on the farm. "Oh," he said, "I never thought about that." Why? Because he was not an agronomist. His field of vision was too narrow.

Perhaps nowhere on the farm is specialization seen more obviously than in confinement factory livestock facilities. These single-use structures enslave multiple generations to a single commodity production: dairy, hogs, eggs, broilers. They harbor pathogens that build up resistance to medications, sanitizers and fumigants.

In contrast to all this, nature is diversified. Perhaps no one has better examined this contrast than Allan Savory, with Holistic Management. His book and lectures provide ample evidence to the failing of assembling a bunch of specialists in one room to analyze a problem, but not using a holistic, everything-relates-to-everything approach during the process.

Healthy landscapes are not monocultures; they are polycultures. To be sure, each organism performs a specialized task, and multiple organisms in proximity is common, but nature's theme is toward diversity, symbiosis and synergism, not the other way around. Each organism exists in balance with others. Imbalances always correct themselves with population adjustments. Foxes get mange; wildebeests starve.

I would like to ask Don Tyson (former president of Tyson Foods, the worlds' largest chicken producer) a question: "How many chickens are enough?" Living in the poultry capital of the east, I believe sometimes the industry really will not be satisfied until every square foot has a poultry house on it. It doesn't matter that they've turned the Shenandoah Valley into a toilet bowl because of the overabundance of poultry manure. After all, this is poultry production country. They just feed it to cattle.

Many of our problems in crop production are due to growing monocultures instead of polycultures. Of course, Wes Jackson is one of the heros in this department. Maintaining large monocultures takes huge amounts of labor and energy, whereas polycultures offer greater resiliency. A specialized farm uses machinery a couple of months a year and then it sits the rest of the time.

Specialists have given us the ability to plant, fertilize and harvest corn using global positioning satellite technology--WOW!--but nobody is asking the question: "Should we plant corn?" As Stan Parsons says in his "Ranching for Profit" seminars: "We've become incredibly accurate at hitting the bull's eye of the wrong target." Food production is fundamentally a natural thing, not an industrial thing. And a natural thing leans toward diversity, not specialization.

SIMPLIFIED VS. COMPLEX

The second pillar of an industrial paradigm is simplification because every job must be doable by any dummy. The goal is to hire cheap labor, take thinking out of the process, and dehumanize everything to minimize variation. That is why fast food joints have a plethora of timers, buzzers, pre-cuts and low-choice menus. They need low-skill staff and that requires taking the brain out.

Voluntary serfs who join vertical integrators in the poultry and swine industry do as they are told by the specialist "field reps." The multinational corporate structure does not lend itself to experimentation based on the farmer's ideas. The only ideas worth adopting are the ones that come down from the appropriate academic departments at accredited government-run institutions.

Perhaps this pillar is epitomized by the famous quote attributed to Henry Ford, father of the modern factory, who allegedly said: "The only thing I don't like about this is that I have to hire a whole man, when all I need are his hands." Doesn't that sound like the kind of guy you'd love to devote your life to? A real lover of humanity, wouldn't you say?

In contrast, nature is complex. Even the most simplified landscapes are a complex network of plants, animals, bacteria, minerals and water. Furthermore, farming requires analyzing a complex number of variables in decision-making, often in a split second. A foolish culture entrusts its food supply to simpletons.

A simpleton accosted me at the stockyard a couple of months ago, wanting to rent a couple of acres in a riparian zone that we had carefully fenced off to protect it from the livestock. I carefully explained that we were trying to protect the stream and wanted to let the vegetation grow up for landscape diversity, wildlife enhancement, and watershed rejuvenation. He replied, exasperated: "Well, you're not using it. Why can't I?" How do you talk to someone like that?

The natural result of this simplification template in agriculture is a monumental brain drain from America's countryside. For several decades now we've exported our brains to the cities to be bankers, lawyers, engineers and scientists while the C and D students stayed home to run the farm. Is it any wonder that the sharp siblings in the city are absconding with the wealth of the countryside? On the whole, farmers are no match for their academic city cousins.

The stereotypical redneck bumbling buffoon out on the farm is ubiquitous: Jon Arbuckle's parents are farmers in the Garfield comic strip; Zero in Beetle Bailey is a farm boy. And now the droves of 40's-something cubicle-sitters yearning for a family-friendly country experience, away from the daily expressway battle, are buying up farmettes thinking this farming thing has to be simple. "Nothing to it. Any old dummy can farm."

The factors that create livestock and plant health, let alone soil health, have occupied the pages of ACRES USA for decades; yet each issue has something new. This world of humus, actinomycetes and omega-3 fatty acids is complex, complex,

complex. It is an interconnected web complex beyond the human imagination, not a simplified factory system of input-output.

ROUTINE VS. FLEXIBLE

The next pillar of industrial principles is routine. The same thing, day after day after day. The material may change but the procedure stays the same. The machine does the same function day after day after day. Workers show up for shifts at the same time every day of the year, for decades.

Routine is necessary because of the huge financial and emotional capitalization required to set up the assembly line. Office cubicles are close to the assembly line concept. The Dilbert comic strip, of course, pans this whole monotonous routine and the bureaucracy involved with maintaining status quo. Every reasonable suggestion meets with derision from higher ups. It's the nature of the beast.

Routine is important to insure that what goes out the back door every day looks exactly alike. In a factory, interchangeable parts won't be identical very long if different machines or procedures are used in the manufacturing process. Identical copper fittings falling into the bag at the end of the assembly line are a product of routine everything.

In contrast, nature is flexible, or dynamic. It changes every day. Floods, droughts, hurricanes, epidemics--the face of nature is ever fresh, ever new. Birth, youth, maturity and finally death are in direct contradiction to the routine. Nature is full of variables, full of changes.

A pond built today looks sterile and unalive. But soon,

overflying birds drop in some seeds and eggs. Migrating amphibians set up housekeeping and in a few short years the pond is full of plants and animals. This natural dynamism is magical almost, happening regardless of human input. Fresh foods deteriorate and eventually decompose, regardless of human manipulation. To create nonperishable qualities in food is to eliminate its nutritive capacity. Living food decays.

This is why a landscape cannot be immobilized, or frozen in time. National parks, with a no-cut, no-burn policy, are accumulating dead and downed biomass at unprecedented rates. The large Yellowstone fires of a few years ago are harbingers of things to come. You cannot freeze a landscape in time.

Food produced seasonally and eaten seasonally, from its own bioregion, is far superior to that trucked in from a thousand miles away. Why must everyone be able to buy fresh tomatoes in January? More important, why in the world would the organic community try to duplicate this pillar of the industrial paradigm?

When we began marketing eggs to restaurants, I offered preventive apologies for the quality reduction that our chefs would see in the winter when the birds cannot be out on fresh daily pasture. One chef responded: "Oh, that's no problem. In chef's school in Switzerland, we had special recipes we used for May eggs, June eggs and July eggs. Every month had a little different nuance in the eggs, based on local layer diet, and we capitalized on the differences with the menu."

Isn't that powerful? In our industrialized U.S. culture we want eggs to be the same color every day of the year. We want sameness, sameness, sameness, instead of celebrating the dynamic changes from nature's pantry. The identical offerings from vending machines perhaps epitomize the unnatural qualities of much American food. The industrial giants extrude, reconstitute, amalgamate and adulterate so that everyone can enjoy soilant green. We want perfectly uniform tomatoes, cardboard apples and french fry potatoes--everything to fit the same box and the same picker.

Life is distinctive, ever-changing and fresh. "Variety is the spice of life" is a saying as old as our culture, and true about life. Food is life. Let's treat it that way.

MECHANIZATION VS. BIOLOGICAL

Here, to me, is the crux of the paradigm difference. Food is animate, not inanimate. Viewing life as a mechanical process creates all sorts of ethical and moral dilemmas.

Every historical culture, from Hindu to Scandinavian, from Native American to Judeo-Christian, has maintained a reverence for life. The mystery surrounding life brings awe and wonder to the heart and mind of any reasonable, non-proud person.

We can take apart a plant to its cells, to its molecules and even to its electrons, protons and neutrons, but we cannot put it back together again. There is a mystery of life that far exceeds the human grasp, and a reverential distance that must be maintained when dealing with life. In our mechanistic view we arrogate divinity to ourselves, and say: "I think the pig should have been made this way."

And so with genetic engineering we make tomatoes that are half pepper, and pigs that are part human, until nothing is distinctive at all, but a jumble of DNA. We say to the chicken "there is nothing special about a beak." It's in the way when we crowd 9 birds in a

16-inch by 22 inch cage, one wire tomb among thousands in a football field-sized factory house. So we cut off the beak.

We put pigs on slatted floors in crates, thousands upon thousands. Nothing is special about the plow on the pig's nose--it is useless. In fact, pigs would be a lot better off without them. Our cleverness to manipulate has no ethical boundary.

The movie Jurassic Park contains the pregnant line, uttered by the journalist to the brilliant scientist intoxicated with his brilliance: "Just because we can, should we?"

Today, we must dare to ask the same question of our industrial food community. Just because we can clone a sheep, splice tomato genes into pepper plants and terminate soybeans, should we? Are there really any ethical questions at stake?

No, not if life is nothing more than a pile of protoplasm. Not if individuality is a misnomer, distinctiveness an old-fashioned idea. A pox on humane animal production! Animals are only piles of electrons, protons and neutrons, suitable for interchanging and manipulation as we see fit. And if that is what life is, why do we wonder that our young people have no self-worth? Why do they say life is cheap, nothing sacred?

At the risk of being declared a raving ecology freak--I am not--let me say unequivocally that biology is not mechanics. Life is not inert. How dare we approach life as if it were just as subject to our discretion as a lump of clay, a hot iron bar, a spinning mass of plastic fiber? The result is mechanical food and mechanical people--lifeless, bored, depressed and devoid of meaning.

At its most foundational level, food is biological, not

mechanical. Wresting life from food leaves it devoid of life-giving properties.

As members of the clean food movement ("industry?" ha!) we must understand the diametrically opposed paradigms regarding food and life between us and the multinational industrial complex. Our passion, not to mention our apologetics, in articulating these differences is key to helping people understand our position.

Here are some simple questions: "Is food animate, or inanimate?" "Are there any ethical boundaries to life manipulation?" "What kind of person would you like producing your food?" These can act as catalysts to a healthy discussion of the underpinnings of a clean food system. If we allow ourselves to get bogged down in the "how" instead of the "why," in the details instead of the conceptual, we will lose every time because data and statistics always carry more wow power, more pizzazz, than philosophy.

Believe me. The soft underbelly of the industrial paradigm is philosophical. Steering conversations down that road may yield some sweet victories in this war over what is on the world's dinner plate.

Industry vs. Biology

Chapter 32

Developing a Sense of Ministry

Families tend to stay together when the cause is big. A family friendly farm is one that exudes a sense of ministry, a sacred calling.

A desire to make money is not a ministry. To carry an enterprise multi-generationally, the business must literally gush with outreach, big picture, missionary worldview.

Perhaps a discussion about marketing will help to lay some groundwork in this discussion. Every marketing guru has a system. I hear folks talking about marketing surveys, making sure you have the right audience, feasibility studies, developing marketing plans, and strategic positioning for market share.

Every permutation on the above theme exists in countless marketing texts, and I've read a pile of them. Each has something to offer. We should all be reading more of these texts.

But in all the scheming and visioning, I think the most foundational element to marketing is to believe it does make a difference how your product is produced. Differentiation is the key to niche marketing, but product uniqueness is only part of the equation.

The other part is that, in your heart of hearts, you must believe that it does make a difference whether beef is raised on grass or in a feedlot, that it does make a difference whether the laying hens can run around on green grass or stay cooped up in a cell on a fecal factory farm.

Too often we get caught up in the mechanics of marketing and forget the "why" of our product. The ripple of influence carries over into our neighborhood, our state, our country and our world. A deep conviction that there is a right way and wrong way, societally-helpful way and a societally-debilitating way, is what fuels our marketing passion into the heart and soul of the buyer.

In a day when we praise mediocrity, and when price is the only bargaining agent, we must understand that in the great conflict of the ages between evil and righteousness we champion what's good. It does make a difference if our communities smell of fecal feedlot particulate. It does matter if our waterways carry effluent from malfunctioning or overflowing manure lagoons.

And the message that must penetrate every junk food junkie is that it does matter how we live, what we eat, and what we spend our money on. Victimhood is alive and well, as is excusehood. Making a difference is what most people yearn to do, leaving a life's legacy that is cherished for multiple generations.

Especially in agriculture, we hear a constant message that eggs are eggs are eggs. Corn is corn is corn. Pigs are pigs are pigs and tomatoes are tomatoes are tomatoes. And yet each of us knows that this is not the case, any more than people are people are people. Some folks we love to be around and others we'd just as soon not see for awhile.

Sameness is the hallmark of industrialization. The interchangeable part requires each piece to carry identical features. We've let this carry over into biology and even politics. We think if we just sat down and had enough committee meetings we'd all come out thinking the same thing. Forget it. Some people really do want bad things. Grass farmers have a positive answer for everything that's wrong with our food system--even fertility depletion on vegetable operations. Good paradigms and bad paradigms do exist in the food production system. The issues are far bigger than personal preference.

We need to get over this notion that we can build a successful full-time farming enterprise out of something that's "neat," or "cute," or even "different." Those mentalities, as a driving force, won't carry us very far. They might sound inoffensive and noble, but they do not generate the energy needed to survive in the great conflict of the ages.

Michael Olson, author of <u>Metrofarm</u>, likens marketing to fighting. "We want to become warriors," he says. While this may not be a politically correct way of describing our attitude toward selling pastured poultry, it is nonetheless the accurate way. You see, I really believe the vertically integrated farming industry is bad for the environment, my neighbors, our economic system, nutrition and our culture. No redeeming value. Just bad, rotten.

So when I produce a clean, nutritious chicken for an appreciative cheerleader patron, the business and its product are not just "neat," "cute" and "different." I've taken a swipe at the evil and won one for righteousness. You say I'm taking this too far? Go out and market something at less than this intensity and see how far you get.

The truly successful farm marketers I know have a deep philosophical conviction about a rightness and wrongness in food production. A belief system that establishes clear lines of demarcation between acceptable and unacceptable. Customers warm up to this attitude, and enjoy participating in this great battle. People love to be involved in great causes. Let's not undersell what we're doing.

Daily we're confronting the unbounded cleverness of the arrogant human mind. We now have genetically modified organisms, offering half human pigs and half tomato peppers. Oh, it sounds sincere enough--solutions for human sickness and suffering, feeding the world (haven't they worn that out yet?), and nutritionally superior food (yeah, right). Irradiation will be called "cold pasteurization" and on and on it goes.

As much as you may believe in making a profit, even that is not enough to drive your marketing. The passion generated by financial gain doesn't hold a candle to the passion created by seeing good triumph over bad.

Yes, it does make a difference. We're not just offering choices to consumers. We're not just experimenting with interesting new models. We are pursuing truth and righteousness, living by conviction rather than expedience. That fuels our marketing. The strategy is secondary. If we could measure our personal commitment to the sacredness of the difference our models make, we would be able to measure our chances of success.

Marketing is only one aspect of this whole ministry idea, but it is a big one. And the way we market fuels this sacredness element. Our children, since they were toddlers, have listened to customers tell us how much their family counted on us for their food. They've listened to people describe skin rashes and respiratory problems due to contaminants in meat, only to have these problems disappear when they began eating ours.

Then they look at our children and tell them: "Your Mommy and Daddy keep our family healthy. And our children are counting on you to keep them healthy. Thank you for what you do." What do you imagine that makes a child think?

Does that make a child think that his life is doomed to oblivion, that the family's vocation has no merit and is just a cog in a big wheel? Of course not. This kind of interchange creates a feeling of doing something noble, of saving the world, and all that stuff. It's a wonderful feeling, and one that encourages the children to stay with it.

Compare that interchange to what the child of an average conventional farmer hears.

- From friends: "Eewww. You work in a smelly old chicken house. Yuck."
- From customers: "You pollute our town's water and air. You should be ashamed."
- From neighbors: "You stink."
- From buyers: "We've already got too much. Go away."
- From the field man: "Disease. Everyone has it. Can't get rid of it."
- From parents: "Ain't no money in farmin'."
- From teachers: "Why don't you go make something of

yourself?"

The average commodity buyer couldn't care less if the farm makes a profit or not. The guy operating the grain elevator doesn't give an extra dime when the crop is scanty.

We've had customers who, upon hearing of a tough season, shove a bill in our pocket with the parting comment: "Look, we depend on you to stay in business so we can eat good food. Here's something to help you over the hump." After they leave, we find a crumpled up $50 bill. And the kids look wide eyed--as parents get teary-eyed--realizing how appreciative this customer is.

We had a patron once call and say: "I'm trading cars. The dealer won't give me anything for this old clunker, and it would make a great farm car. I'm either going to give it to the Salvation Army or to you. Besides, you deserve to drive a better car than you do. What do you say?"

I said: "Bring it down, and we'll take a look. But we don't really need one." At the time, we were driving Teresa's grandmother's 20-year-old Dodge and it suited us just fine. No reason to change.

So the patron brought the car down and it was a black Lincoln Towncar. Teresa and I looked at each other, got in, and took it for a little test drive. When we got back to the house, Teresa said: "Where do I sign?"

We drove that car for two years, and it was one stylish carriage. You should have seen the looks from the neighbors, especially when I told them a customer gave it to us. They were used to the comments itemized above. They didn't know what appreciative buyers were.

The mentality in mainline conventional agriculture is that the farmer grows it because he's supposed to feed the world--the old guilt complex again--and then takes it to market and should be satisfied with whatever he gets. This wholesale marketing paradigm has gradually reduced the farmer's portion of the retail dollar from about 35 cents in the 1960s to about 9 cents today, and it's still trending down. More money goes to the paper company for the cardboard cereal box than goes to the farmer for the raw ingredients.

Do you realize that we could decide, as public policy, to just quit giving the farmer any money at all, and it would only change the price at the supermarket about 10 percent? Now do you see why the average farmer feels beat down and unappreciated? Those feelings do not encourage the children to stay with the farm. And those feelings don't generate happy, contented family times.

Environmentalists have brought something to our culture that was sorely lacking--a true appreciation for the physical world. The Romantic poets certainly planted the seeds for the kind of musings that graced John Muir's writings. Of course, we must guard against the pendulum swinging so far in over-correction that we become tree worshipers. But when God created it He said it was "very good." And Wendell Berry eloquently defends the sacredness of the physical in his essay "The Gift of Good Land."

God didn't just lead the Israelites into a piece of real estate. That real estate, with its carefully defined borders, was a place. A physical place. And it did matter that they not cut the fruit trees. It did matter whether or not they rested the ground every seven years. It did matter how they treated the animals. Each procedure was carefully articulated in the divine law. Unfortunately, for most of our western culture's history, the dominance theme has prevailed to the exclusion of the nurturing theme. Because the physical is evil and worldly, it carries no bearing on "how shall we then live."

The Amish tradition carries forward the notion of interconnectedness between the physical life and the spiritual. How we farm, how we milk the cows, what we eat--all these reflect on the spiritual condition. The physical and spiritual create a whole, not two unrelated compartments. Our farm, from its physical layout to the substances produced on it, is nothing more than a manifestation of our spiritual condition. Likewise, the way we view what we produce, the way we view land stewardship, bears directly on our spiritual condition. A dominating land use attitude will produce an arrogant, dominating spiritual attitude with little appreciation for other denominations, for example.

At the risk of being labeled a heretic, I would suggest that defending forage-based cattle production is no different than defending the virgin birth. Why is it spiritual to defend the virgin birth, which establishes the efficacy of Jesus Christ, and not spiritual to defend the creation, which establishes the perfection and order of God? Let me give an example.

Our local pregnancy help center has an annual fundraiser banquet and a friend invited us to go a couple of years ago. Make no mistake about it, we believe in the sanctity of life. Anyone who has helped a cow deliver a calf knows that critter is not just some fetal tissue. When you stick your arm in there, it pulls its legs away from you, rolls around and responds in all sorts of ways. It's very much alive. Nothing about popping into the air makes that calf any more alive than it is when you're trying to get the obstetric chains wrapped around its hooves.

I called the banquet chairman and offered to donate chickens for the meal. "No, it's too difficult to deal with whole birds. Besides, we just get the cheapest thing available," he said.

I thought: "What a joke. Here's a group dedicated to the sanctity of life and they couldn't care less that this chicken is 9 percent fecal soup, that these birds were raised in the most inhumane conditions imaginable, and that the farmers and processors are viewed as cogs in a corporate wheel. Everything about this food is dehumanizing, and yet we're supposed to go and feel all warm and fuzzy about saving babies. Give me a break."

Talk about hypocrisy. With all due respect, the environmentalists certainly aren't immune from these same problems. Plenty of environmental banquets offer Spam, artificial lemonade and cold cuts that represent everything anti-environmental. They just don't put quite the spiritual twist on it that the pro-lifers do. I know an environmentalist leader who has been to a couple of White House lawn bill-signing ceremonies. He told me that he lives on whatever is available in vending machines because he doesn't have time to eat real food.

What? Vending machine food represents the bottom end of the food system, has more packaging material than anything else, and is as anti-nutritional as anything you can find. I wonder how much evil petroleum he's consuming in all the plastic packaging eating out of the vending machines. But he's an environmental poster child, darling of the tree huggers.

A ministry, and its worldview transfer to the next generation, requires a degree of consistency, both in practice and thought. Not complete consistency. None of us achieves complete consistency. Not even the Amish. But we should at least come close, and not participate in these kinds of glaring inconsistencies. Every model we use, every change we make, is adopted to put the farm closer to an Edenic orientation.

We believe strongly that there is no difference between the world we see and the one we don't. Everything has a spiritual dimension. Everything. I wish I knew how many churchgoers feed their cows chicken manure. I wish I knew how many churchgoers foul the water with chemicals. I wish I knew how many churchgoers put their cows on concrete and destroy their ligaments in 24 months. And all the while they send money to missions and speak great platitudes about loving their neighbor--while their fecal factory farm carries a fecal particulate cloud 5 miles downwind.

The family friendly farm must have a ministry view. Ministering to people who eat our food, so that we wouldn't do anything to harm one of these little tykes we've just taken out to hold and pet baby chickens. The industrial paradigm hires Philadelphia lawyers on retainers as a veil of protection between them and litigious consumers. Lacing their pseudo-food with food coloring, preservatives and artificial everything, they have absolutely no "ministry" worldview except more dollars for their stockholders.

What's wrong with giving our kids a big worldview? Putting a spiritual dimension on every activity creates a sense of urgency and a feeling of worth. Nothing can be more healthy than inculcating in them early that the way they weed the beans reflects how they love what's good. Keeping the chickens in water and feed is just as important as defending righteousness and truth. Nothing makes a mockery of God faster than taking a flippant approach to His creatures. Or His trees. Or His ponds.

Hanging above my desk is one of the first cross-stitch creations Teresa ever gave me:

> "No greater purpose has any man
> Than to tend the herd or till the sod,
> Yet leave behind him, richer still,
> Those acres leased by him,
> From God."

That describes our philosophy pretty well. As much as I defend property rights and believe they are essential for proper stewardship, I also don't appreciate how this freedom has been used to exonerate every sort of despicable rape imaginable on the gift of good land. We don't really own the land; it's been here longer than we have and will be here long after we're gone. To create a healthier, more productive landscape and leave that to subsequent generations, related or not, is the least of our responsibilities.

It does matter which trees we cut. It does matter what we sell our customers. Does physical food have nothing to do with spiritual food? When we drink a cup of cold water, does it matter if it has poisons in it? Certainly compartmentalizing life is convenient. It absolves us of having to think through what we're doing. We can participate in our narrow little activist ministries and feel all comfy and cozy about saving a creature or a sinner. And as long as the thinking that propelled us to adopt that cause doesn't penetrate other areas of our life, we can live a blissful, shallow existence.

This came home to me one time when I was asked to attend a high level think-tank meeting in Washington D.C. involving top officials with the premier sustainable agriculture organization at the time and USDA and Department of Health and Human Services.

The goal was to create a program depicting farmers as the real doctors. Societal health starts in the soil. That was the idea, and the attendees began assessing the weak links in the system that impeded consumers from getting good food.

After some opening comments, they opened the floor for ideas to put on the day's agenda. Understand, this was a group of about 15 very important people. I'm not using names because several of them are household names. I was the only farmer there. Pretty typical of these groups--throw the sop out. Let's have our token farmer.

I jumped in with what I considered the weak link--government involvement in the system. I said: "Look, the reason people don't get good food is not because it can't be profitably and ecologically grown. And it's not because there's no market. The growers are here. The buyers are here. But between the two exist a host of anti-entrepreneurial regulations, from employment rules to inane safety codes to food safety and inspection regulations that deny my ability to sell a glass of raw milk to my neighbor."

A healthy interchange ensued, in which I explained the impediments to commerce that the government has erected that preclude producers and consumers from being able to interact freely. One of my friends here in Virginia tried to jump through all the hoops to slaughter and sell his own beef on his own farm. He finally called me in dismay because it would cost $300,000 in order to sell one pound of hamburger. He had to give it up.

Finally, the lady right across the table from me, who was the Washington-based political lobbyist for this sustainable agriculture organization stood up and said: "Look, you're in the country. I'm in the city. I don't have time to get in touch with my food supply."

I was so taken aback, I was speechless. She was supposed to be my friend. I would have expected as much from the government officials. But she was supposed to understand sustainable farming issues and be on the side that would make it easy for people to get cleaner food than the fraternized, colluded government fare. At that point, the discussion broke down and I shut up--for the rest of the day. It was over. She was too busy talking on cell phones and tooling around in her Lexus to get in touch with her food supply. She was too busy lobbying for government grants, too busy saving the amphibians and fighting for clean air, to mother her own children or prepare her own food. I shouldn't assume too much here, but I've seen this kind of smallness time and time and time again, and I get tired of it.

I'll bet she had time to go get in touch with Thumper and Bambi on her most recent vacation, and I'd imagine she's in touch daily with her big screen television set. I'll bet she had time to drink adult beverages with senators and congressmen. If the truth be known, we do have time for the things we value. And we all make choices.

We need to understand that all of us suffer from a sense of smallness. But it is suffering. We should try to rise above the smallness, to think bigger, broader, and more consistently. Incorporating into every fiber of our being the action that comes from realizing everything does really matter engenders deep within a sacredness to what we do. It offers a sanity in an insane world. A sense of purpose in a drifting, anchor-less world.

One other area where I deal with this, and we'll close this chapter. I do a lot of speaking to civic groups like Rotary, Kiwanis,

Key Clubs, Ruritans, Garden Clubs and retiree groups like AARP and Retired Teachers Associations. These folks routinely decry the loss of pastoral landscapes and the filth of factory farms. They complain about the cardboard quality food they buy. And yet these grandmas and grandpas take their little kiddos out for fast food and trinkets. When will we begin matching life and lip? When will our walk mimic our talk?

That's just the problem, we're all talk and no walk. We're all spiritual and no physical. We're all show and no substance. We're all doctrine and no action. We're all platitude and no practice. True ministry, true mission, permeates deep within our souls, and extends into the core world view of our children.

And when we create an agricultural paradigm with this nobility of purpose, with this life-changing outreach, we impassion our children to a daily dedicated ministry. And that gives them energy, direction, goals. It converts a rock-along, ho-hum life into something with eternal value and meaning. Anything less handicaps our children with mediocrity. I beg you to unleash the power of ministry in your family and its business. If you do, you will discover the essence, the real secret, of a multi-generational family friendly farm.

Chapter 33

Business Charity

The family friendly farm is more than business: it is also charity. Our family has routinely done things that didn't make straight dollars and cents sense, but they were the charitable thing to do. At the risk of sounding self-righteous, I want to include some of these things to help people understand how the successful family enterprise is a conduit for charity. Too often in our culture we believe charity is only doled out in self-addressed-stamped envelopes. We vehemently disagree. Charity starts at home, and with our family business.

We had a dependable customer who failed to arrive on time to pick up her chickens so I called her to make sure everything was okay. That is standard procedure when people fail to make their appointments. She answered the phone and her voice began to break. Her husband had just walked out on her and the two children, and she didn't have time or money to get the chickens. She said she was simply too embarrassed to call and cancel the order.

The next day, we took the chickens and gave them to her. Seldom is such a direct-help opportunity thrust into our laps, and I think it's important to jump on these, to seize the moment. As our children played with hers, she just wept and poured out her heart. It was a time of encouragement and healing for her. At that time, seeing unconditional love was exactly what she needed. It was a powerful evening, for Teresa and I, as well as the children.

Another time we had a good customer whose mother died and left a complex estate situation. The mother had lived clear across the country from this customer and the multi-month settlement process really took its toll on this family. She called to cancel her chickens, explaining that she just couldn't make it with all the extra problems in the estate.

We set a date and took the chickens to her. I'll never forget that visit. It was big old ramshackle house situated on a little cleared knoll. The outbuildings were neat and the yard was tidy. She and her husband had bought the acreage about five years previously, and she explained what it was like at the time. They hauled dump truck loads of debris out of the house. In fact, the house was not visible because it was completely enshrouded in multiflora roses, sumac, honeysuckle and young trees.

Abandoned horse-drawn farm machinery, cars, wagon chassis, metal tanks and other cast-offs literally covered the yard and what was now a small adjacent pasture. They were still working on the house, but it was quite livable and the kitchen shelves were lined with her dried herbs. Her homemade herb tea sweetened with their own honey on that cold day warmed my innards. But not as much as thrilling to the sheer effort that the family had put into restoring the abandoned farmstead. She just couldn't believe that we would drive nearly two hours one way just to bring her chickens. Another memorable moment.

Charity doesn't always repay as sweetly. I was driving back

one evening from a bull hunting trip and picked up a fellow hitchhiking. He was down on his luck and had nowhere in particular to go. You guessed it, a classic early homeless. I dropped him off at the shelter in town. After making sure he had a place to sleep that evening, I gave him my card and said if he ever needed anything, to give me a call.

About two days later he called, and asked to take me up on the deal. At the time, Teresa and I were just feeling like this farm thing would actually fly. We'd been at it for about seven years and it was finally turning the corner from hand-to-mouth to actually a little left over at the end of the year. I went in, picked him up and brought him out for the day. He was an excellent worker, and especially good around the children, who were small at the time. Extremely polite, he struck us all as a guy who really could benefit from a hand up--not a hand out.

We couldn't afford to pay him, and we didn't really need the help, but we were willing to make sure he had a place to sleep and plenty of food and clothes. We talked about fixing up an outbuilding and he quit smoking. Either Teresa or I drove into town every morning and picked him up and took him back to the shelter each evening. We paid him enough for some spending money, but we were investing a fair amount of time in the two round trips to town every day.

He withstood the pressure from the director at the shelter to quit coming out. The director told him this was how wealthy farmers attained their life of luxury--abusing poor people like him. The fellow said he tried to explain that we were making some plans but it was useless. He was happy; I was happy. It all seemed very up-and-up.

Things went along well for two weeks. Then we needed a

two-week break because my brother and family were coming for a two-week visit and we decided it best not to complicate that family time with this fellow. I arranged with a friend to provide employment during the two week period. All systems seemed good and we felt like maybe we could really help this fellow get back on his feet.

He didn't go to this friend's business the first day. I called the homeless fellow and he said he'd fallen from a ladder and had to go to the hospital. No record of this at the shelter. I called the hospital. No record of anyone by that name. I called him back. He said some unrepeatables and hung up. Never heard from him again. Kind of a sad tale, but maybe another day, another fellow will yield better results. Our children were very cognizant of what we were trying to do, and still remember him, even though they were quite small.

From time to time we've had extra eggs and we've always taken them to this shelter. We buy our work shirts at the thrift store there. Charity is always tough in these circumstances because it's hard to determine what a person actually needs. The worst thing to give a lazy person is a handout. Many folks don't need money, they need discipline and budgeting advice. How to help someone in a way that is meaningful requires wisdom and discernment.

When I was working at the newspaper, one of the reporters dressed up in some old clothes, grew his beard scruffy, and decided to try his hand at panhandling downtown. You must understand, Staunton, Virginia is a very conservative town of 20,000. In the newsroom we all contrived the sad tale and he got it down pat after a few attempts. We refined it here and there until it was really touching. Then he headed for the street.

He spent one whole day--an eight-hour stint--plying his

trade. He eluded police by moving around from corner to corner. He finally got picked up by the police toward the end of the day, but only after he'd accumulated nearly $100. Of course, the newspaper editor eventually explained the situation and the police released him. The money went to some local charity. The point of the story, though, was that Americans have a real soft heart, and this can be turned against us by shrewd operators.

In our family, though, we believe it's better to be taken advantage of once in awhile, than to have a hard heart.

A couple of times we've had quite a few extra eggs around Christmas because the restaurants we service really slow down at that time. It's hard to regulate the production of a chicken to perfectly match the consumption. When that happens, we give eggs to businesses we patronize, and it does wonders for public relations. We receive preferential treatment the rest of the year. It's amazing how little charity it takes to make an impression. We take these eggs to the bank, farm store, tire store, small engine mechanic, and whoever else takes care of us throughout the year.

We give all of our neighbors chickens or a Thanksgiving turkey each year, but that cannot be considered pure charity. These farmers don't take too kindly to the traffic our customers create on this dirt road when they're trying to get round bales hauled home, or running silage wagons back and forth. City people are supposed to stay in the city and country people are supposed to stay home. That's their mentality.

You know what they say about people who live on a dirt road. We never have to lock our car doors except in August--to keep the neighbors from putting runaway zucchini squash in them.

Anyway, our gifts to neighbors help keep us in good graces.

Business Charity

The situations where neighbors get crossed up could fill a library. Especially if you're the new kid on the block, you need to be sensitive to how your presence may be changing things in the community. Often, it's just plain ignorance of how country folks interact. The art of rural communication is not mastered in a day, nor a year. It slowly grows on you, kind of like a dialect. Expectations about fences, barking dogs, roaming cats, snow plowing, road passing and a host of other topics can be as different as east and west when local and foreigner meet. Be the first one to give, and you'll find helpful, charitable neighbors. That is, give everything but advice. Be charitable with everything but advice, and you'll get along just fine.

Much of our giving today is just trying to maintain an open door for information. You wouldn't believe how many letters I get in the mail with a dozen questions, and not even a stamped, self-addressed envelope. We have stopped giving free, escorted farm tours. But often, at night, I receive numerous phone calls wanting advice. Am I complaining? No, but I think it's important to understand that most successful businesses do not get there by holding their things close to their chest.

We've loaned tools and had them come back broken. Loaned vehicles, and they come back broken. Given things away, only to be accused it wasn't enough. You don't build a successful business without a few bounced checks and false accusations. That's what happens when you deal with people. But that is the exception, not the rule.

Look for opportunities to be charitable and especially for the business to do things that do not return money. This creates a healthy spirit in the family. The charity always comes back, and proves the truth: "It's more blessed to give than to receive."

Multi-Generational Transfer

Chapter 34

Retirement: An Alternative View

"WE'RE SPENDING OUR KIDS' INHERITANCE"

This bumper sticker adorns the rear of many a Winnebago. Retirees tool from RV park to RV park, taking in Elderhostels and souvenir traps along the way. The souvenir traps, of course, are for the grandchildren. "Grandma and Grandpa were here, and we thought of you," they tell their family on a stop-by.

The jet-setting, world-traveling retiree is a scourge on western culture. At the very time when a person has the wisdom and experience to really shepherd succeeding generations through the political, economic and social quagmire of the times, he's completely out of touch with his family. Is it any wonder that our culture does not respect the elderly when their ambition is to experience a second childhood? They are carefree, attending plays and musical events, with seemingly no responsibilities.

Of course, the other end of the teeter totter is equally wrong--full-time child care. Plenty of grandparents are surrogate parents for career-worshiping, materialistic children who want everything their parents have right now. Shame on young families who foist upon the grandparents daily childcare responsibilities. And shame on grandparents for agreeing to do it. Some day, those kids need to grow up. The greatest thing you could ever do for these egocentric kids is pull off their umbilical cord. Don't aid and abet their jaundiced perceptions any longer--help them grow up, and you'll be doing everyone a great service.

When you sit back and reflect on the world's most stable communities, you find a close business, social and economic interweaving of the elderly, the middle-aged and the young. You do not see the elderly pushed off into their own village or community. You see each generation integrated into the preceding one.

Retirement was invented in the 1930s - a politically expedient short-term way to clear older workers out of factories and make room for the unemployed younger workers. But like so many government solutions, once implemented, and once they've outlived their usefulness, they develop into entitlement programs.

When social security first began, 40 workers supported one recipient. In 2000 the ratio is down to 3 to 1 and trending lower. As the baby boomers reach retirement age, some say the ratio will drop to about 1 to 1. Who thinks this will endear the elderly to the young?

To be sure, the role of the elderly changes, as it should. As we age, we can't keep up the physical pace. But a society which has no interconnected role for the elderly is a sick, sick society. The western culture's love affair with euthanasia, nursing homes and assisted living centers, or age segregated villages, is sick, sick, sick. It is just one more manifestation of how compartmentalized our society has become. It starts with church nursery, daycare and kindergarten, and ends in nursing home. The more we can age segregate society, the more no one has to deal with the needs of the other group.

In the one-room school, the older children helped the younger children. "You never know a subject until you have to teach it" is a ditty all of us repeat, but what middle-school child in today's world ever tutors a second grader? I'll never forget a high school senior who rode the school bus with me when I was in second grade. Her name was Lenora, she had beautiful red hair, and to me she was the most classy, elegant lady I'd ever met. For some reason, she took an interest in me. She would let me read her literature book aloud to her as we rode along, taught me how to do long division, and talked to me like I was a human.

That year, I entered the reading contest at school and won second place. That was the first event of what became an interscholastic and then intercollegiate career in forensics, drama, journalism and debate. She was truly from a bygone era of mutual respect, before it was normal for older kids to look down on younger kids as scumbags and punks. I remember her more keenly than any of my elementary school teachers except my first grade teacher, who was also a neighbor and one of the most enthusiastic, passionate, kind people ever to walk the earth. When I think back on those dimly-remembered school days, Lenora is the one who invariably comes to mind. She was 18; I was 8. We had 30 minutes in the morning and 30 minutes in the afternoon. But what an impression she made.

Retirement. It's the scourge of a society in which our

vocation is not a lifelong passion, but a means to play. What a shame that we have trivialized work in our culture to the point that what motivates people is not the work, but the off work time. We work in order to afford a vacation or a bigger screen TV for weekend entertainment. We work to get a pension in order to play our retired years away.

Work is a noble thing, the defining characteristic of the human. Good work requires thought, long-term planning and a sense of mission. The whole notion of retirement is a creation of the industrial age, where work is outside the home, outside the soul, and only a card in the time clock. We have bastardized work and glorified play, the inevitable consequence of trivial, meaningless vocations and an egocentric culture.

I've been blessed to know just a couple men who have not retired. They just keep going. Invariably they are self-employed businessmen, and invariably they are revered by their middle-aged children, their neighbors and their business associates. Extremely successful, they certainly do not work as hard physically as they once did, but the fire of passion still lights their eyes as they anticipate each days' challenges. They're also fun to be around because they're not living in a dreamworld.

Because they're still matching wits with today's problems, their wisdom is priceless. I respect their perspective because they stay up-to-date with trends. They read the newspaper for the news, not for the entertainment section. They invariably have an optimistic outlook and a zest for living. And that's the way I want to age.

Retirement, as practiced in our culture, destroys the value that normally comes with age. If only retirees could see how they've cheapened their value by dropping out of society and voting for more perks at the public trough. No wonder they are becoming more resented and more despised. To be honest, I don't know exactly how a typically retired person ought to interact with family and what ought to occupy his day if the adult children are all at the workstation and the grandchildren are in school.

This whole problem of what to do with the elderly is simply an extension of what to do with anyone except the most highly productive 25-50-year-old class. Children are seen as a liability because they contribute no economic benefit to the family. The elderly are seen as a liability for the same reason. It is tragic that our culture has lopped off the two ends of the age spectrum that most need each other for establishing roots. The world wide web may have us more connected to the branches and leaves, but we are more disconnected from our roots. It takes both to grow a majestic old tree.

In family friendly farming, retirement is a word that should not exist. Instead, the older folks incorporate themselves into the heartbeat of the farm in meaningful ways. As patriarch and matriarch, they offer a multi-year perspective.

Let me tell you about my maternal grandmother, who for me provided a pattern for what I'm trying to describe. I never knew her husband, my grandfather. He was an alcoholic and it was a tough life. She was already widowed in my earliest memory, and when I was about 14 or 15, she came to live with us on the farm. She purchased a small mobile home and moved it right outside our yard fence.

Although she had social security, she never drove a car or traveled. I'd be mowing the lawn and she would come out to give me a big glass of iced tea or lemonade. She ate dinner with us a couple of nights a week. Mom usually took her to town if she had to shop, and the rest of us took care of plumbing and other things for her.

She would routinely fix supper for us if we were busy in the hayfield or some other big project. It was very much a give and take. She certainly depended on us, and that is as it should be, but we also depended on her. She crocheted afghans, made fancy doilies for the sofa and cross-stitched. Mending work clothes for us was a specialty, and she enjoyed doing it.

During my high school years, one night she called me on the phone: "Could you come over? I think something's in my house."

I ran over and she said she thought it was in the kitchen. Hearing nothing, I finally pulled open the top drawers where she kept the silverware. Imagine my surprise when looking right back at me was a skunk! I slammed the drawer.

"Grandma, it's a skunk!" I can still remember her laughing.

"Well, get him, by cracky."

I got a broom handle and reopened the top drawer. As I opened it, he jumped into the next drawer down. I shut the top one and opened the second one, but the skunk was already heading downhill again. I chased him down to the bottom drawer and when I opened it, he jumped back up into the third drawer. The skunk seemed to be playing games with me.

We went up and down the drawers a couple of times. Then I got smart and skipped one. Sure enough, he was there, and ready. Instead of his face, he had his rear gun well aimed, cocked, and ready to fire. I felt the wet spray hit my face. Grandma maintained a safe distance and suddenly yelled: "There he goes!"

The skunk leaped out of the drawer and ran across the floor. I whacked him with the broom handle just enough to stun him. Then I kicked him across the room, opened the door and kicked him out. Grandma scurried around opening windows and doors. Fortunately, it was not winter. I ran home and took a long, long shower. Believe it or not, I never did stink. But Grandma's house sure did. She moved in with us for about a month until the odor abated.

After high school I was often home here alone or with my sister, working. Dad worked off the farm at his accounting and Mom taught school. My older brother was already in college. Grandma would often fix lunch or supper for us, giving us exactly what we wanted. She didn't try to turn it into a "new food" lesson; she just fixed what she knew we enjoyed.

She watched her baseball games--being from Ohio, the Cincinnati Reds had all her affection. She knew the standings and could give play by play accounts. I never figured out how she could crochet afghans and keep up with the ball game, but she did. I can't ever remember her sermonizing or pontificating. Oh, she wouldn't back away from a political discussion, and she could give first person accounts of the depression. Born in 1899, she lived through fascinating days in history.

After Teresa and I were married and had Daniel, she held her great-grandson as if he was the most special guy in the world. And I suppose he was. One winter she went up to visit some relatives in Ohio and slipped on a doorjamb, falling and breaking her neck. As she went into surgery, the doctor explained the seriousness of her condition. She was frail and old and it was risky at best. She just smiled at him and said: "It doesn't matter. Either way it goes, I'll be fine." She never came out of it, dying a couple of days later. I saw family members I didn't even know in those next few days.

But I don't think any of them knew her like I did. She had always been a frail woman, having had three husbands, went through a couple of divorces, but here, on the farm, surrounded by loving family, mutually dependent, I believe she came into a sense of purpose, an abiding faith, and the satisfaction of knowing that, because of the grandchildren surrounding her, her life had not been in vain. I cherish her memory, her wonderful sense of humor and her zest for life.

When our cattle shademobile blew over the first time, it was sitting out in the meadow where she could have easily seen it from her house. This is a 50 foot x 20 foot portable loafing shelter. We had it jacked up kitty-corner getting a tire fixed. During that two-day period, a heavy windstorm came through and flipped it over. Grandma was really disappointed, but not in the fact, only that she hadn't seen it go. "Boy, I wish I'd seen that," she lamented. She was always ready for excitement.

Here's my point about this story. My memories and the impact she had on me, my appreciation for that period in history and all the subconscious roots and elderly affirmation she offered me could never have happened if she had been tooling around in a Recreational Vehicle enjoying her second childhood. Her dependency on us kept her close and required that she focus her attention on home and hearth.

The physical, emotional and economic independence of retirement tempts the elderly away from family and home. "But what would I do? Just vegetate?" asks the retiree. Here's a little list for Grandma, just for starters:

- Fix a meal for your adult children and their kids once a week.
- Take your grandchildren on walks.
- Help your daughter or daughter-in-law can food.
- Help the young women in the church fellowship can food. Many of these young women don't know how to process food and don't realize the economic savings it offers the household.
- Tutor a neighbor child or grandchild. You can tutor any subject in the early years.
- Bring cups of iced tea to your grandchild who's mowing your lawn.
- Help in the garden.
- Host friends and business associates who visit the family home business.
- Teach grandchildren how to sew, crochet, and knit.
- Be aware of what your adult children and grandchildren are doing.
- Read, read, read to your grandchildren.

How about a list for grandpa?

- Prune the grapevines, and show your grandchild how to do it.
- Build model rockets, cars or whatever with grandchildren.
- Help in the garden.
- Help in maintenance, fixing things or building things.
- Help other young men in the church fellowship with their projects.
- Read, read, read to your grandchildren.
- Read contemporary stuff to stay sharp and provide cultural leadership to your clan (dare we say "tribe?")
- Enjoy recreation with the grandchildren--fishing, hunting, bowling, shooting, etc. Don't go by yourself; take them along.
- Take the grandchildren for long walks.

I know retirement has not eliminated all these things. I've watched retirees do all this stuff, and admirably. Plenty of retirees still do these things and endear themselves to the next couple of generations in wonderful ways. But I fear that too often our retirement system has made it too easy to neglect these things. Or, most commonly, these things are done sporadically rather than routinely. We're back to the "quantity time" vs. "quality time" myth. In reality, there's only time. Period. And you can't build a relationship with the next generation by entering their lives once a month or during holidays. It takes immersion.

A family friendly farm offers that immersion like nothing else does. The grandparents are perfectly suited to indulge the clumsiness of a child during the learning years. This takes the pressure for indulging the little child's slowness off the high-performance parents. Each generation can fill its most efficient niche within the business.

We believe in taking care of our own. The older generation must realize it can't practice an independent spirit and then suddenly expect the adult children to respond to dependency. Mutual dependency is far stronger than independence because none of us is good at everything. We get much farther by using others' strengths where we are weak.

But now comes the kicker. How does the older generation instill within the succeeding generation a notion that our possessions are really just things and best used for the needs of others? We don't instill that notion by seeing how much we can amass so we can live comfortably in our old age. "Saving for retirement" becomes the be all and end all of wealth accumulation.

Rather, as we age we should be giving the children our wealth. As we meter it out to them, they can get farther and will have the means--economically and emotionally--to care for us graciously as we age. Too often farm families go along with the older generation amassing wealth, amassing wealth, amassing wealth while the young people really put the energy, creativity and spirit into keeping the farm going. Then all of a sudden, at some point, the whole kit and caboodle lands in the lap of the next generation.

In his wonderfully insightful book QUIT TODAY PAY CASH DON'T RETIRE AND MOST IMPORTANT DIE BROKE Stephen Pollan points out that $10,000 given to a 25-year-old can start a business. But that $10,000 given at 50, through inheritance, will probably pay for a two-week European fling. The gift is the same, but the timing creates the worth.

The resentment over having worked slavishly while the parents raked off the cream can't be undone. And suddenly acquiring wealth does not temper the gift with years of regulated responsibility. You can always tell when this happens. The business goes through a giant hiccup, and many do not survive. A few years after Dad died, a couple neighbors remarked to me that they were amazed how the farm never bobbled. Things just kept getting better. From the outside, nobody could even see a blip in the screen.

I have a couple of neighbors that have weathered this transition just as gracefully. In each case, I had the privilege of watching the dads and their adult children work together. It was beautiful. And the things that those dads said about their boys to me in private--oh, it was special. Nothing but admiration and praise. But the wealth and the decision making of the farm was divested early and given over bit by bit so that when that fateful day came, the dad was working for the children and not the other way around. When the transition occurs like that, it is like a beautifully choreographed dance.

Then there are other farmers who say: "If I gave responsibility and assets to my kids, they'd lose it or sell it right out from under me." I've got news for you, parents. If that's the way you feel, you may as well sell it now and enjoy the money, because it won't change suddenly when you die. On the other hand, perhaps the reason you feel that way is because you've never tried it. You might just be surprised. It may be that your 45-year-old kids are chafing under a father who can't relinquish any control.

Estate planning and business books tout the need to rapidly escalate your income during the 40s and 50s in order to get that retirement nest egg together. After all, the reasoning goes, when your income producing potential wanes, you'll need every bit of that to keep you out of the gutter. My question is: "Where are the kids in all this scurrying, dashing wealth accumulation?"

Let me offer a different perspective. What if, instead of getting together $250,000 in equity for retirement, you forgot about that and lived humbly through your 40s and 50s, letting the adult children have it all? Then when you hit 70 years old, instead of having all this money, you have grateful, hardworking children who have leveraged that money with their energy and your experience to grow the business?

The money pumped back into the profitable business via incentives to the young will generate far more equity than you can by keeping the children as slaves and dependents as long as possible so your accumulation at the top can be higher. My point is this: if I didn't think my kids would do whatever was necessary to take care of me if I were injured or became ill, I'd consider myself a failure as a parent.

That may sound harsh, but what kind of parent enters middle-age fearful that their children will let them end up in the gutter? Even most dysfunctional parent/child relationships would not end up like that. All of us are investing in something. My question is, in real value, where is the most payback? Is it investing in our personal nest egg so we can enjoy a second childhood, or investing in the children so they will fall over themselves taking care of us when we age?

I know this flies in the face of estate planners. It flies in the face of our culture. But you can't live for yourself and then expect the kids to think first of you. If our actions indicate we need to take care of ourselves first, our children will grow up with the same idea. But if our actions indicate that we really want others to succeed, really want others to reach their potential, then those people we've enabled will be emotionally, if not economically, indebted to us and will rise to the occasion when our needs are great.

A couple of times we've needed money and we've just let a couple of customers know about the need. "How much do you need--$5,000, $10,000?" The reason is because our customers know we work for them. We do not charge what the market will bear. We charge what it takes for us to live well and enjoy self-respect. We often give things away to customers. Better to charge top dollar for the premium stuff and give away the dregs than drop the price on the premium and never be able to give anything away.

"But what if we need that money? What if the kids make some erroneous decisions?" The older folks can come up with all sorts of faithless questions. Can you not trust your kids? If you can't, then you have problems a lot deeper than you realize. I would hate to age independent of someone who would care for me. The Golden Rule comes into play here.

We have a couple of elderly firewood customers who are unable to split or stack their wood anymore. When we take their load, we split it and stack it all up nice for them. They've asked me:

"Why do you do this for nothing?"

I respond: "Because when I get like you, I hope a young crew of guys will do this for me." Folks, this is just the way we ought to live. This is not my religion over your religion. This is not Wall Street. It's about living today with the end in view, building real equity in tomorrow. From our earliest days, Daniel has helped me take care of these older people. He knows about the need to care. And I am grateful for his knowledge.

Teresa and Rachel have taken care of her grandmother, cleaned the house, done laundry. Your kids should grow up watching you take care of the elderly. This engenders in them a life rule to care for those unable to care for themselves.

In this whole retirement and elderly care arena, our society has lost a lot of ground. It hasn't developed overnight and it won't heal overnight. Tragically, many of the elderly, growing up in that independent American spirit, will not allow their children or grandchildren to care for them. And many whose children do try to care for them don't appreciate it, are grouchy and whiny, and make life miserable for their family care givers. Bad attitudes on anyone's part should not be tolerated. Families should all band together, in solidarity, to help the old grouch get rid of this terrible habit.

Again, I'm not an estate planner or a financial wizard. But I do believe a family friendly model does exist, and offer some of these musings to stimulate additional thought and discussion. If I'm still around in 20 years I may change some ideas, but right now I'm betting on Daniel and Rachel. And I'm blessed to have two such incredible performers to bet on. And I truly believe that if more parents would take the attitude Teresa and I have, they'd be surprised at the emotional and economic returns this paradigm offers.

Teresa and I have no desire to be rich. I don't think the kids have any desire to be rich. We want a legacy and a life. Rather than putting money away for us, we make big capital improvements in the farm in areas where the kids are interested. We'll ask the kids: "What big project should we do?" I'm not spending this for me; I'm spending this for them, and for their children.

Who cares if the bank account is small? As the farm offers more viable opportunities for more family members, that surely is as wise an investment as a stock portfolio in Fortune 500 companies. Dad was an accountant, and a good one. I remember well when he went through his high income time.

He and Mom paid for the farm in 10 years. Then they invested in a rental property because he didn't think the farm could turn as much money as quickly. At that time, with our knowledge level, he was right. Now I know differently, but that's getting ahead of the story. Through high school I remember many days working at that old three-unit apartment house in town. I also remember getting calls from the police at 3:00 a.m. because someone there was drunk and causing a ruckus.

One guy apparently ran a drug ring out of the basement apartment. He tore the place up when he moved out, too. Two single girls moved into the top apartment. They never took out their garbage. Finally Dad and I had to go in and carry a months' worth of rat-infested, half-decomposed garbage out to the curb. Yes, we evicted them, but by that time, their pets had urinated all over the hardwood floors. We had to rent sanders and completely refinish the floors in order to get the smell out. Finally he sold the place. He made good money, but got tired of the people problems.

The farm still wasn't a viable entity, so he and Mom began investing in the stock market and precious metals. I well remember one morning he and I were out fixing some fence. When he came in, the stock broker called and said he'd been trying all morning to get him because this one stock, if he'd sold at 9:00 a.m. would have netted him a cool $5,000, but now, three hours later, it lost $5,000. Dad looked at me and decided to get out of that rat race. Thank goodness we didn't have cell phones then. We'd have been entangled even deeper.

At that point, he and Mom made a definite investment shift. They decided to invest in something that they knew and something they could control. That was the farm. He paid for a survey, which was very expensive with 9 miles of perimeter. But I remember how I felt when that happened. Suddenly he was investing in my life. Invigorated afresh, I tackled the farm work and its long-term planning with renewed energy. It was a major turning point in our relationship.

I felt like he had come home and had quit chasing rainbows. I've never forgotten how I felt in those ensuing months. I felt freedom, long-term commitment, and like his heart really was here. Our conversations went to a new level and I had a much deeper appreciation for who he was and what he wanted to do. His dreams became mine. I appropriated to my life his vision. And it all happened when I really felt like his heart was here on the farm with the family business.

I think older men, facing their energy reduction, often get derailed right when it's most critical to stay with the plan. I think Dad, frustrated at not being able to make the farm go full-time, began looking at these other endeavors as the new opportunity. In fact, it was during this time that he seriously considered selling the farm when he found out how much it had appreciated in 20 years. But the frustration over the apartment and the off-farm investing brought him home. I'll never forget when he said: "From now on, I'm investing here."

True, those investments did not give immediate monetary dividends, but they paved the way for me to devote my young energy, unconditionally, to the farm. And he was able to age surrounded by grandchildren and the fields and forests he loved. My appreciation and devotion to his vision grew enormously during that time, and neither of us ever looked back.

May I encourage you, parents, to bless your children with this feeling? If you can't sign over the deed, at least sign it over emotionally. In turn, you will enjoy your children's adoration, respect and care. Now that's a portfolio I covet. And it probably best expresses the family friendly farm.

Chapter 35

Inheritance: Performance Distribution

Inheritance is a wonderful institution, as old as humanity. And yet this is probably the most ticklish facet of all family multi-generational businesses. It's especially critical in farms, where the equity is tied up in non-liquid assets. "Land rich and cash poor" is common enough to be axiomatic.

The first aspect I want to deal with is what I call "the sin of equal inheritance." Let's go ahead, drop the bombshell, and get this one out of the way right off the top. In western cultures, the most common inheritance practice is to give equal portions to each child. In fact, in the unlikely event that disparate portions are bestowed, the small-end beneficiaries resent being loved less. In other cultures, this is not the case. The Israelite culture, for example, gave the firstborn substantially
more than the other siblings.

To examine this issue further, I think one of Jesus' parables sets the stage as well as anything. It's the parable of the talents, given in Matthew 25:14-30, quoted here from the King James Version:

"*For the kingdom of heaven is as a man traveling into a far country, who called his own servants, and delivered unto them his goods. And unto one he gave five talents, to another two, and to another one; to every man according to his several ability; and straightway took his journey.*

Then he that had received the five talents went and traded with the same, and made them other five talents. And likewise he that had received two, he also gained other two. But he that had received one went and digged in the earth, and hid his lord's money.

And after a long time the lord of those servants cometh, and reckoned with them. And so he that had received five talents came and brought other five talents, saying, Lord, thou deliveredst unto me five talents: behold, I have gained beside them five talents more. His lord said unto him, Well done, thou good and faithful servant: thou hast been faithful over a few things, I will make thee ruler over many things: enter thou into the joy of thy lord.

He also that had received two talents came and said, Lord, thou deliveredst unto me two talents: behold, I have gained two other talents beside them. His lord said unto him, Well done, good and faithful servant; thou has been faithful over a few things, I will make thee ruler over many things: enter thou into the joy of thy lord.

Then he which had received the one talent came and said, Lord, I knew thee, that thou art an hard man, reaping where thou has not sown, and gathering where thou has not strawed: And I was afraid, and went and hid thy talent in the earth: lo, there thou hast that is thine. His lord answered and said unto him, Thou wicked and slothful servant, thou knewest that I reap where I sowed not, and gather where I have not strawed: Thou oughtest therefore to have put my money to the exchangers and then at my coming I should have received mine own with usury.

Take therefore the talent from him, and give it unto him which hath ten talents. For unto every one that hath shall be given, and he shall have abundance: but from him that hath not shall be taken away even that which he hath. And cast ye the unprofitable servant into outer darkness: there shall be weeping and gnashing of teeth."

Several observations are in order:

- The master knew the abilities of his servants. He had given them enough freedom, and had worked with them long enough, to evaluate their track records. He could predict, with great certainty, the way each would use his talent.

- The master, in this case, Jesus, did not love any of these servants any more than the other. Each was equally loved. The distribution was based strictly on propensity to be productive, or what we would call performance.

- Even the high risk one got something. The master continued to give opportunities for even the unproductive servant to prove he wasn't negligent.

- But the master had a breaking point, a day when all records were called in; a day when mercy ended. And on that day, the unproductive servant lost even what he had.

- The master is just. Period. Whatever you may think about this parable, the master is absolutely just. In our politically correct day, it's important to understand that we must not question the lord's fairness. It is a given; assumed.

- The servant who received two talents did not complain about the one who received five, nor did he complain after everything doubled.

- Expecting performance is proper and reasonable. Satisfaction with status quo is not a commendable thing.

- It is not selfish to expect our assets to increase in value. It is good stewardship. Having no expectations is slothful and wrong.

Now back to the topic of inheritance and the sin of equal inheritance. Let me tread into unspeakably personal waters. In our own family, the three children took decidedly different paths. My older brother moved away, married, and went to the mission field as a bush pilot-mechanic. My younger sister moved away, married and went to Texas. I stayed home and loved this farm.

Dad and Mom's love for all three of us did not change with any of these decisions. They supported my brother and his family financially and emotionally. My sister married an accountant and the family has been financially successful. These life decisions did not mean anything as far as character, love, good or evil were concerned. All of us are different, and we chose different lifestyles, different vocations. That's fine.

Our family did not sit here at home and castigate my sister for moving to Texas. We did not sit around and bemoan my brother's family being on the other side of the world. Nobody thought ill of anybody for making these decisions.

But when Dad and Mom made their wills--and we talked about this, by the way--it didn't take a rocket scientist to realize which child had the ability, the love and the track record to make this farm perform. I got the farm and the other two got money, car and moveable assets. Does this mean the other two were loved less? Not at all. Is this unfair? Not at all. Did this mean I got more value in inheritance than the others? Yes.

If you want to hear something unfair, let me describe what is unfair. What is unfair is when a couple of children leave the farm, go to town, drive around in fancy cars, go on vacation cruises, enjoy paid medical insurance and a 401(k) while one sibling stays home on the farm, doing with less, shopping in thrift stores, and after investing his whole life on the home farm to keep it going, has to pay off the other siblings in order to keep it. That, folks, is the sin of equal inheritance.

Nobody held a gun to the heads of the other siblings and said they had to leave. Each chose their life path. And it is the height of unfairness to saddle the child who puts his blood, sweat and tears into the farm with a payoff of the siblings who invested their lives in another course. If you think I begrudge the non-farming siblings their vacations, paid medical and fancy cars, you've missed everything. I think they deserve those things. They worked for them and they earned them. Those items represent accumulating value based on the investment of their life's energy.

But thousands upon thousands of farm boys come to their 50s and 60s, having invested their life's energy in a farm, only to lose it to their non-farm siblings because Mom and Dad were afraid if they gave the farming child more of the farm, the others would curse them forever. This makes my face red with anger. It makes my blood boil. I wish I could reach out of these pages and grab every elderly farmer who is getting ready to disrespect his stay-at-home son or daughter with payments to the non-farming siblings. I would shake him until his teeth rattled.

Like the teacher shaking the little boy after he'd misbehaved. She said: "Johnny, I think the Devil's got aholt of you."

The boy replied: "I think he has too."

Many of these non-farm siblings, when they visit, do not even take enough interest to walk around the barns or fields. They visit on special occasions, eat the food and enjoy the farmhouse ambiance, then head back to their steady paychecks and paid vacations. Then they have the audacity, the unmitigated gall, to think they deserve equal inheritance. How dare they. How dare anybody be this greedy, this selfish, this unloving? Few things make me more indignant than seeing how middle-aged farming children are treated by their non-farming siblings during an inheritance settlement.

They say estate settlement brings out the worst in people, and I think it does. I'm sure someone will say all this is easy for me because I happen to be the one who "got mine" in this case. Just a minute. Teresa's family has a farm. She has three brothers. We have told her father over and over that we do not want anything if one of those guys wants the farm. We have no desire to take money from that farm when we've invested our lives here and those guys have invested theirs there.

Who stayed here and held this place together when taxes doubled? Who fixed the roof on the house? Who fixed the barn roof when it collapsed under a heavy snow? Who mowed the fields and made the hay? Who stayed up with calves at midnight, slept with chicks during a rat infestation, and kept the line fences up? Who took care of Dad when he was ill, washed him when he lost control of his bowels, and took care of the farm so he wouldn't have to worry about the work getting done? Who mows Mom's yard and takes her to the hospital at 3:00 a.m.?

Farms don't just stay viable on their own accord. They don't just stay like those pleasant childhood memories without continued hard work. Rot, rust and depreciation take their toll. Were it not for the child who stayed home and cared for the farm, that festive farmhouse ambiance that all the grown children and the grandchildren enjoy would not exist. It would be gone. Mom and Dad would be in a condo and the farm sold or grown up in brush and in complete disrepair. It would be a place of unsightliness and poor productivity.

The farm child's equity is tied up in the land, buildings and farm machinery. Typically, the farm child doesn't have paid vacations, fancy cars, 401(k) plans and an RV as physical manifestations of his life's work. Instead, his is seen in the healthy cows, lush pastures and tight fences.

The sin of equal inheritance has destroyed far more family farms than multinational corporations, bankers and government regulations. When the estate is divided equally, several things normally happen.

- All the siblings can agree, in perfect harmony and in one accord, to put it in a trust with the farmer sibling in charge. The non-farm siblings agree to forgive interest, or to an interest scale that will not break the farming family. This is highly unusual. Most families have at least one black sheep.

- The non-farm siblings can just forgive their portions. Realizing that their parents were not thinking about the best interests of the farm, they can rectify the situation post-testate and return it to the way it ought to be. This would be a miracle. Same problem as the first one.

- Since the siblings can't agree on the disposition, and since usually at least one sibling wants his share in cash, the farm is sold and the money split equally among the heirs. This moves the farm out of the family. That family farm is over.

- The land is surveyed and the children hire an appraiser to make sure that each heir gets an equally valuable piece of the real estate. One child gets the house and less land while others get more land but no improvements. This is fairly common, but the problem is that it reduces the farm to a less viable full-time farming land base.

- The heirs can wrangle for years, pay huge attorney's fees, part with a lot of ill will and continue tolerating each other but deeply resent each other for the rest of their lives, passing on the same feeling to their children, ad infinitum. This is the most common.

Wouldn't it just make a lot more sense to call a family conference, while everyone is still alive and in their right mind, and hash out the disposition of the farm? Let everyone say their piece and come to a consensus so that everyone knows exactly what will happen? For some reason, this subject is taboo in families. The estate disposition is some deep, dark secret. Why can't we all appreciate our finite existence, talk about the end and deal with the thorny issues in order to keep them from being so thorny? I will never understand why the inheritance issue is feared. It may not be the most pleasant topic, but it's inevitable. It is the one family farm occurrence that is inevitable.

Obviously, the disposition I've described has a few prerequisites. First, at least one child must have a track record of performance. Obviously, if none of the children has a track record, then all bets are off. This whole discussion is moot. Divide it however you want. And if the child who stayed has a bad track record, maybe none of it should be left to him. Division when no clear performer is around is not the issue here, and I don't have suggestions for it either, except as I deal with parents who have no children. All the farms with no clear front runner provide opportunities for the folks who want to get into farming. There's plenty to go around. I just hate to see the farm children disrespected.

At this point it would probably be important to establish that I don't think it is the parents' responsibility to establish any child in any business. If parents have the means to provide financial assistance, that's fine. But no child should assume that his parents owe him a special opportunity dispensation. The whole theme of this inheritance discussion so far is to get parents to not destroy a deserving child, and to not destroy the family farm in the process. To not recognize the contribution of the farm child is as inappropriate as children assuming they deserve parental goodies. Historically, parents have helped their children get established, and that is how family wealth gradually accumulates over time. But no child should enter adulthood assuming that his parents will front end his ventures.

I can hear you saying that giving the farm child the farm is certainly an unfair advantage. Perhaps it is if you view a farm as nothing more than an economic enterprise--just business. But I view the farm, and specifically the family farm, as a cultural icon, as a foundation block of land stewardship, healthy communities, nutritious food and a cultural heritage. Does anyone really want unskilled, low-paid foreigners and multi-national corporations controlling all the land and being responsible for all the food? Does anyone think this would provide the greatest care of either the land or the food supply?

I believe the family farm merits special consideration. I do not believe it needs government subsidies or any government programs. The way to preserve family farms is for individual farmers and their families to think creatively. That means marketing differently, producing different things, including entertainment, and shaking up the conventional estate planning protocol. None of this takes an act of Congress. None of it takes political lobbying. None of it takes any politician doing anything. All it takes is for each of us involved in agriculture to think differently. That doesn't cost the taxpayer anything. It's something you and I have control over. It's at least a starting point.

I am not an estate planner. I believe in using good estate planners, attorneys who specialize in writing wills, and accountants. Here are a couple of things that can help transfer the estate tax-free:

- Divide the estate between Mom and Dad. As soon as my parents decided to leave me the farm real estate, they split the farm 50/50 so it would come in two pieces. That way the estate tax ceiling would double. At the time, it was $600,000, and this technique raised the nontaxable value to $1.2 million. The obvious risk of such an arrangement is divorce or disagreement. If one spouse decides he wants his half of the farm in cash, the other one has to cough it up from somewhere. These things are never pure victories. But real breakthroughs usually involve some risk.

- Gift $10,000 per year and keep notes payable against the estate. This is very effective if you start early enough. When added to the technique above, it allows $20,000 in value to be transferred annually. The tax-free gift ceiling is $10,000 annually. You can receive as much as you want, you just can't give any one person more than $10,000 per year. What you do is gift $10,000 to the farm child. Then the farm child loans you $10,000 and holds the note payable. These notes accumulate year after year. When the parents die, these are presented as outstanding bills against the estate just like the utility bill or the car payment. If the amount is high enough, it essentially transfers the estate in clear settlement of the debt.

Wills do change over time. For example, Teresa and I have a living trust with guardians for the children and trustees named. This will is good until Rachel hits 21. At that point, we will redo the will to reflect the independent decision-making capability of both children. Daniel has already established himself as a farm performer, but we want to give Rachel equal dibs. If she marries and leaves the farm, we will update the will to reflect the change. If she stays, it won't need much change.

What about buy-outs? Obviously, one of my goals is to see farms generate enough income that the stay-at-home child could actually just buy out the parents. Or in an equal inheritance situation, the farm income could buy out the other interests. This has certainly been done and may be more viable in the future if land prices fall. But I think it's important to understand that just because something has been done a certain way for a long time does not mean it is still viable today.

Mom and Dad bought this farm in 1961--550 acres, a pole barn, a pole equipment shed, house, a couple pieces of hay equipment--for $49,000. At that time, feeder calves brought about $30-$35 cwt. and you could buy a brand new tractor for $2,000. In the 40 years since, the value of this farm has gone up to $500,000, which is an increase of roughly 1,000 percent, or a ten-fold increase. Feeder calves sell for about $70 cwt., which is only a 100 percent increase, or a doubling. A new tractor sells for $30,000, and that's not a big one. That increase is about 1,500 percent, or a fifteen-fold increase.

In the 1940s and 1950s people interested in farming could routinely capitalize the farm and pay their salary on the farm-generated production. In those days, it took roughly one cow's worth of land to produce one cow. Today, it takes 10 cow's worth of land to produce one cow. I wish this were simpler to describe, but can you see what's happened? The commodity price has risen slightly, but the land, machinery and inputs have risen astronomically. And this disparity is causing some of the foremost agriculture economists to preach a new sermon. The new sermon is that owning the land is a completely separate business than farming.

Allan Nation, editor of Stockman Grass Farmer magazine, says that he sees a division coming between land ownership and land control. No longer will people who own the land be the ones farming it. Because of this widening gap between land value and the value of what is produced on the land, he sees land ownership as a defensive posture adopted by those who have already made their money, and land rent as an offensive posture adopted by those who want to earn money.

For example, we can rent all the $3,000 per acre pasture we want for $20 per acre per year. That acre will produce 200 cow-days, or about 350-400 pounds of beef per acre, which works out to a value of about $250. You cowmen, please don't crucify me for being off on some of these numbers. They're ballpark figures for the sake of discussion. If we can generate $250 per acre at a land rent cost of only $20, the economics obviously works out better than if we generate that same $250 at a land ownership cost of $250 (assuming a 30-year mortgage at 10 percent, factored for inflation). I know these are rough figures, but they illustrate why the disparity between land and land production value is fundamentally changing the way multi-generational transfer occurs.

Just because grandfather did it this way 50 years ago doesn't mean it's a viable solution for today. Obviously it would be nice for the farm to generate so much cash that the second generation could just buy out the previous generation. And I wouldn't suggest that is impossible. I'm only suggesting that it is not likely, and not really a viable solution in the real world.

This is exactly why so many young farmers who inherit the right of first option to buy the place--parents consider that their respectful plum to the farm child--go belly up 10 years later. The falling commodity prices simply won't keep up with the interest. What normally happens is that the farm child takes an off-farm job in order to generate the cash to keep making the payments and pay the taxes. The farm is essentially a nonpaying second vocation.

To be perfectly honest, some places do still exist where this disparity is not as large and where land can still be purchased with on-farm production. These areas tend to be in the rural areas with declining population. Especially attractive are cotton belt areas. The worst areas are in the far west, where rainfall further limits production, but where population incursions hold land prices up. In our economy, land prices for the most part bear no relationship with the productive capacity. And except for main transportation routes or around cities, this is a fairly new phenomenon.

Teresa's grandfather, who did not go beyond the eighth grade, made enough money on his little beef cattle farm in the late 1940s to 1970 to take the extra and buy a couple of rental properties in town. That was a different economy. I heard an older large-scale farmer recently lamenting that when he started, he could take a truckload of cattle to the sale barn and trade it for the biggest new tractor available. Today, the same load of cattle would scarcely buy the tires for that tractor.

This is what is really hurting the Amish, where the tradition is that the parents essentially loan the children money to buy their farms and then the children gradually pay it back. This income supports the parents in their older years instead of pensions and social security. The child who stayed on the home farm is expected to pay the value of that farm back as well. As long as land prices remain relatively stable, this is a workable option. But in 20 years, these families have watched their land values rise from $500 per acre to $5,000 per acre.

One Amish friend sold out and moved, although I wrestled through the decision with him for a year. He bought his farm for under $1,000 per acre but the county put a public water and sewer line through it to supply a large industry about one mile away. When he received his real estate assessment, the farm was valued at $10,000 per acre. What would you do? He sold out and moved to Tennessee, where he could still buy land for $1,000. I can't fault him.

I know the tendency is for the older generation to say: "Look, I did it. Why can't you? You must just not want it bad enough. This young generation is soft. Get moving. That's what I did." But we live in a different world, a different economy. Forty years ago you could trade a cow for an acre. Today it takes ten cows. That's not just a matter of working hard. The same deal can't be done today, and it's insane--and mean--for the older generation to expect it.

And that is why we must talk openly and honestly about this whole inheritance issue. I watch these young farmers worrying about what they're going to have to do, completely in the dark as to what Mom and Dad are planning. I watch the older parents just rocking along expecting their adult children to run blindly ahead, not questioning anything. The need of the hour is for families to sit down and talk it out. Usually, when everyone can come to the table secure in the knowledge that they can present their perspective honestly and without retribution, amazingly creative solutions rise to the fore.

Now the final question: "But Mom and Dad won't talk about it. I'm here running the farm. My siblings are all gone. I'm 30 years old. I've invested now a hefty portion of my life here and I'm doing a really good job, if I must say so myself. Is there a future here for me? Am I going to have to come up with $500,000 for my siblings when Mom and Dad pass on? I don't know anything. I work alongside Dad every day. Yeah, he's slowing down, but he still pulls his load every day. Boy, I don't even want to think of not having him. But my 5-year-old Juniorette loves the calves, does a great job helping with water and feed, carrying grapevine prunings and such. She loves the farm. But will it be here for her? What am I supposed to do?"

Now the rubber hits the road. My advice, based on countless conversations like this, is very simple: "Demand that your parents talk to you about the will. If they refuse, tell them you're gone."

Oh, but that sounds harsh. It sounds unloving. No, my friend. What's unloving is Mom and Dad sitting there all pious and self-assured thinking their child should happily break his back to keep the farm going, and all the while they're plotting to hammer him with a half-million dollar debt when he's 50 and his daughter is right ready to make her career plans. Don't give me this jazz about sounding harsh.

Look, if you're going to lose the farm at 50, you might as well walk away from it at 30 while you still have energy to start over. At 50, you'll just cash it in. Life is too short to spend it wondering, wondering, wondering. If you are 30 and you've done a super job on the farm, plenty of opportunities exist to start over. We had an apprentice looking for an opportunity. He put an ad in Stockman Grass Farmer and had eight requests in 24 hours. I wish I had a nickel for every person who has come by here and said: "I've got a farm but no experience. I need a passionate young person to take it and run with it. Do you know anybody who could do this?"

Like I said above, we are rapidly moving into the time when land owners will be well-heeled, defensive investors. Tomorrow's big "getting in" opportunities will probably be partnerships between farmer entrepreneurial managers and equity-strong older landowners. And that's not a terrible deal. Remember, in 1800 only about 5,000 people owned all the land in England. In 1900, it was owned by 1.5 million. Things do change. That means for every scenario that closes shut, another one opens. The trick is to be informed and up on the trend so you can spot those new opportunities before they become the rage.

If I thought for a minute that I would have to scrounge together hundreds of thousands of dollars to pay off my siblings, I'd leave tomorrow. They didn't live on $300 per month for the first seven years of their marriages. They didn't drive $50 cars. And I bear them no ill will, at all, for their decisions. But because Teresa and I sacrificed, Mom and Dad were able to age here, enjoy our children here, and see their legacy successfully stewarded to the next generation.

And if your parents cannot see the importance of talking about the arrangements now, they never will. Oh, they'll try all sorts of emotional blackmail. They'll say things like: "If you really loved us, you'd . . . " Or maybe they'll say: "That's none of your business. Our parents never discussed it with us." I don't know what they'll say, but I'm sure if this is a taboo subject, they won't cotton to your asking. But you may as well know something sooner than later.

It's not worth the family and marital trauma. I've watched young farm children absolutely die inside, wondering and worrying about what their parents were up to. Get it resolved today, and if it can't be resolved, give it a resolution deadline and then pull the plug. Don't put yourself, your children, or your spouse through the emotional grief of a horrible surprise, and the mental wrangling over unknown scenarios.

Bottom line: talk about the inheritance. Recognize farm economics. Reward stewardship. Follow the parable of the talents. Most of all, realize that inheritance can be the most beautiful thing or it can be the most tragic. It's in your hands.

Chapter 36

But We Don't Have Children

Childless farming couples are quite common today. Many of these couples took part in the back to the land movement but also bought into the overpopulation and scarey world myths. As a result, they chose not to have children and now are too far down the age curve to change it.

An equally common situation is for a farming couple with children to not have a single one who really wants to farm. The children are grown and gone and Mom and Dad are trying to keep the farm going, but it's getting more difficult.

Plenty of options exist for these couples. They can sell their farms once they can't keep up and go live somewhere else. They can sell the land and retain the house, or build a smaller house on a corner of the property.

I'm not going to spend time dealing with all the possibilities they have to divest themselves of the farm, including giving it to charity. Staying with the multi-generational theme of this book, I want to deal with the next step.

But We Don't Have Children

People have often asked me at conferences: "This is all well and good for you to talk about because you're kids both love the farm. But what if neither of them took an interest in it? You can't guarantee that, you know."

Here's my answer: "I'd be actively pursuing someone to inherit it to."

"What? You mean you wouldn't give your kids anything?"

"No. If they won't steward it, why give it to them? I want it to go to someone who will love it and make it perform." And as much as I love my kids, and as honestly as anyone can know their heart, that is exactly what I would do.

This is the balance to my scenario where the farm child goes to Mom and Dad and requires that they discuss the will. I'm sure that hits some people as incredibly greedy and aggressive. I just see it as asking for a return on investment. But in this case, Mom and Dad go to the children and say: "Okay, we're 45 years old. You guys are heading into careers. We'd love for one of you to take the farm. If one of you wants it, we'll make sure you can have it. But if none of you comes forward within the next year to start putting together a track record, we're going to pursue someone else to inherit it to."

If everyone in this situation would do that, it would sure create some wonderful opportunities for farm loving kids who didn't grow up on one. While the pool of young people wanting to farm is small, they are out there. We run one year apprenticeships and these are wonderful young people who would make a farm couple proud.

But we don't search for them on the Internet or advertise for them. We prefer Christian homeschoolers. We prefer no college, although there again we've had exceptions to the rule. Things we are watching for are:

- **Laziness**. When you're here so we can meet and check each other out, you'd better be up when I am, period.

- **Passion**. I want to see a fire in your eyes. I want an eager beaver look, a hunger for this farming vocation.

- **Aggressiveness**. When we walk in the field, stay with me. If you can stay one step behind, you could stay right with me if you wanted to. Get with it.

- **Motivation**. Do you see where you can help and jump in, or do you stand there until I ask for help?

- **Clean cut**. We want the classic all-American boy look. No dyed hair, no earrings, no baggy pants or clothes worn hip (backwards cap, untucked shirt, untied shoes). No matted hair or ponytails.

- **Appreciative attitude**. Being here is a privilege, not a right. We'll respect you and treat you like our own, and we expect the same.

Where to look for such a person? Almost every community now has a homeschool support group. The thing we notice about homeschoolers is that they are not peer dependent and most have a fabulous work ethic. They haven't spent their young years sitting behind a desk. Doing chores and helping around the home are carefully reinforced activities.

I suggest you advertise in publications that have our farming bent. You can find some of these listed in the resource section at the end of this book. You should attend alternative agriculture conferences. I am a firm believer that when the student is ready, the teacher will appear. If you'll walk in circles where these kind of people are apt to be, the two of you will get together.

Often a relative, even if he's distant, can be offered the opportunity. These high quality young people exist. The real need is not to create these young people, but for childless middle-aged farming couples to appreciate their finiteness and get started right away going through what will be a multi-year process at best.

By all means do not assume that because a person has an agriculture degree, he's qualified to be a farmer. More often than not, that experience will assure failure. Don't be afraid of that vocational student that can't master calculus or physics. Often that youngster will shine both in the shop and with living things.

If we really care about what happens to the niche of God's creation entrusted to our care for one generation, then we should also care about it for tomorrow's generation. That means stepping clear outside the bounds of acceptable protocol to find a surrogate heir, a deserving next-generationer who can benefit from a head start. We praise head starts in everything else in our culture, why not praise head start farming?

Don't be put off by this chapter if you don't have someone in mind right at this very moment to carry on your farming legacy. Someone, just that special someone, is waiting to carry on where you left off. Give them that opportunity. And young person, serve your mentor well. Be a worthy son or daughter. Neither oldie nor newby will ever regret forming this partnership. Even with the weeds--and there will be a few--the roses will blossom. And what an adornment roses would be on farming landscapes.

Summary

In the depth of your soul, what is valuable to you? Money? Fame? Prestige?

I've read bookshelves full of biographies and self-help business manuals. Once the rich and famous reach the twilight of their careers and muse about soul-level wishes, their thoughts invariably turn toward home and hearth. The way we treat those who share our bedrooms and bathrooms becomes the defining factor of our life.

The musings of the elderly rich and famous contain confessions, regrets, and apologies for not being there for the spouse and the kids. For missing important events in their children's lives. And too often, for sacrificing the family on the altar of success as defined by Wall Street

Most world famous people, from evangelists to psychologists to talk show hosts to environmentalists, leave a legacy of broken marriages and dysfunctional children. How many, in their twilight years, would trade their money and importance for the honor, respect, admiration and healthy passion of a spouse and grandchildren? I'm not suggesting that riches and fame inherently preclude family success, but statistics certainly point that way.

So why are we all caught up in having the latest gadget? A late-model car? A house bigger than average? Kids dressed in designer clothes? Every time we grasp for those Wall Street success measures, we sacrifice values that are more lasting and more important. The simple truth is that we can almost never have it all. The only way to have it all, if that is to be our lot in life, is to put marriages, family, and integrity first These are the ultimate expressions of our religious convictions and our core values.

A business simply manifests to the community what the owner believes. A farm stinking up the neighborhood, poisoning earthworms, and disrespecting animals or plants presents a certain belief system to the community. A farmer with passionate, well-adjusted children taking over the family business illustrates a certain active belief system. By the same token, the farmer who advises his children to leave the farm because it has no future also illustrates a certain belief system.

Our family, our business, and our landscape are simply extensions of what we really value. Don't tell me what you believe. The world is full of talkers. Don't tell me about your ability to mediate conflict; show me by staying married to your spouse. Otherwise don't pass yourself off as an expert in conflict resolution. I am always amazed at the number of people who go to divorced marriage counselors.

This is why Paul's epistles in the New Testament clearly detail the family life of elders, as a physical manifestation of a belief system. He wasn't nearly as interested in talk as walk. Don't tell me you love your children when you spoil them with trinkets and candy and they are less behaved than your pet dog.

This world is full of phonies. To be sure, we all make mistakes. But mistakes are expected and easily forgiven. Phonies are a different story. Farmers who say they love animals and then coop them up in factory confinement concentration camps are phonies. Fathers who say they love their families but don't read to their kids several hours per week are phonies.

Husbands who say they love their wives but don't leave them love notes or get them flowers occasionally are phonies. Mothers who say they love their children but hire surrogates to raise them are phonies. Forget the money. I have little patience for whiners who can't make ends meet but spend $2,000 a year on clothes.

Our accountant services numerous clients who earn 50 times more than we do. Each year they sit before him shaking their heads, despondently lamenting: "I just don't see how we're going to make it." Their outgo always catches up to their income.

Our family has developed a little ending for the cliche: "I don't have the money." Folks use that phrase as a cop-out for all sorts of things. We believe it would be more honest to add a phrase at the end: "for that." Somehow people seem to get the money and time for the things that are really important We can buy that country club membership, but somehow can' t afford to take a lower-paying job that would allow us to be home with the family every evening.

Time is the same way. We can fit in that movie or that show, but we don't have time to play catch with the kids. I'm guilty. We're all guilty of inappropriate prioritizing in our lives. Children are incredibly forgiving of mistakes. They aren't forgiving of phonies.

Most of us choose our own shackles. We choose our own shortcomings. Goodness knows, the shortcoming buffet offers a diversified portfolio. Want a better marriage? Work on it. Want a better relationship with your kids? Work on it. Want a change of lifestyle? Work on it.

The changes might not happen all at once, but the cumulative effect of many things done right will culminate in success. Remember that the opposite of success is not failure, but depression. Successful people fail just as much as unsuccessful; they just get up one more time. And nothing is more empty than being successful at the wrong things.

What do you want to be successful in more than anything else in the world? Your religion? Your marriage? Your children? How tragic that most success is in the realm of fame and fortune, with the family tossed like a carcass on the trash heap of life. Being successful at the wrong thing is worse than not being successful at all.

What you are most successful in portrays to the world that which you value most. I have no desire to build a farm empire - - even if it is environmentally friendly. Empires aren't environmentally friendly. Some people think I should travel all the time and speak about sustainable agriculture. It's tempting, but tempting like any selfish vice.

No, I'll be content to wear thrift store shirts, drive 10-year-old cars, and devote myself to healing this little niche of God's creation entrusted to our care, loving my first-and-only bride, and impassioning my children with a zest for life. At the end of my days, may it be said: "He loved his God. He loved his family. He loved his farm."

How about you? What will you do? I pray that your investments will be in the portfolio that cannot be touched by the vagaries of Wall Street. I encourage you to find a ministry where you and your family can leave our culture richer for your having passed this way. None of us knows how long we will have to build such honest equity. But whether it's one day or 80 years, this equity bears interest beyond the wildest dreams. Go for it.

Appendix A

Resources

Periodicals

Acres USA
P.O. Box 91299
Austin, Texas 78709
1-800-355-5313

American Small Farm Magazine
267 Broad Street
Westerville, OH 43081
(614) 895-3755

APPPA Grit
American Pastured Poultry Producers Association
5207 70th Street
Chippewa Falls, WI 54729
(715) 723-2293

Countryside & Small Stock Journal
W11564 Hwy 64
Withee, WI 54498
(715) 785-7979

Growing Without Schooling
2380 Massachusetts Ave Suite 104
Cambridge, MA 02140-1226
1-888-925-9298

Holistic Management Quarterly
1010 Tijeras NW
Albuquerque, NM 87102
(505) 842-5252

Homeschool Digest
P.O. Box 374
Covert, MI 49043

The Permaculture Activist
P.O. Box 1209
Black Mountain, NC 28711
(828) 669-6336

Quit You Like Men
152 Maple Lane
Harriman, TN 37748
1-800-311-2263

Small Farm Today
3903 W Ridge Trail Rd
Clark, MO 65243-9525
1-800-633-2535

Small Farmer's Journal
P.O.Box 1627
Sisters, OR 97759
(503) 549-2064

The Stockman Grass Farmer
282 Commerce Park Drive
Ridgeland, MS 39157
1-800-748-9808

Books

Barker, Joel Arthur
**Paradigms: The Business of Discovering the Future*

Bartholomew, Mel
**Square Foot Gardening*

Beckwith, Harry
**Selling the Invisible: A Field Guide to Modern Marketing*

Berry, Wendell
**The Gift of Good Land*
**The Unsettling of America: Culture & Agriculture*

Boyer, Rick
**The Socialization Trap*

Bromfield, Louis
**Out of the Earth*
**Pleasant Valley*

Byczynski, Lynn
**The Flower Farmer : An Organic Grower's Guide to Raising and Selling Cut Flowers*

Carnegie, Dale
**How to Win Friends and Influence People*
**How to Stop Worrying and Start Living*

Celente, Gerald
**Trends 2000*

Cobleigh, Rolfe
**Handy Farm Devices and How to Make Them*

Coleman, Eliot
**Four-Season Harvest*
**The New Organic Grower*

Cook, Glen Charles
**500 More Things to Make for Farm & Home*

Covey, Stephen
**The 7 Habits of Highly Effective People*

Damerow, Gail
**Chicken Health Handbook*

Davidson, James Dale
The Sovereign Individual

Davis, Adelle
Let's Cook it Right

Doane, D. Howard
Vertical Diversification

Dobson, James C.
Dare to Discipline

Douglas, J. Sholto
Forest Farming

Drucker, Peter F.
Managing in a Time of Great Change

Ekarius, Carol
Small-Scale Livestock Farming : A Grass-Based Approach for Health, Sustainability, and Profit

Ensminger, M.E.
Beef Cattle Science

Faulkner, Edward H.
A Second Look
Plowman's Folly
Soil Development

Ferguson, John
Farm Forestry

Forbes, Reginald
Woodlands for Profit and Pleasure

Fryer, Lee
The Bio-Gardener's Bible

Gatto, John Taylor
Dumbing Us Down: The Hidden Curriculum of Compulsory Schooling
A Different Kind of Teacher: Solving the Crisis of American Schooling

Gibson, Eric
Sell What You Sow
The New Farmers' Market

Groh, Trauger
Farms of Tomorrow Revisited: Community Supported Farms/Farm Supported Communities

Hensel, Julius
Bread from Stones

Howard, Sir Albert
An Agricultural Testament

Ishee, Jeff
Dynamic Farmers Marketing

Jackson, Wes
Altars of Unhewn Stone
Becoming Native to This Place

Koenig, Neil
You Can't Fire Me, I'm Your Father

Leatherbarrow, Margaret
Gold in the Grass

Lee, Andy
Chicken Tractor
Backyard Market Gardening

Levinson, Jay Conrad
Guerilla Marketing Handbook

Lindegger, Max O.
The Best of Permaculture

Logsdon, Gene
The Contrary Farmer

Macher, Ron
Making Your Small Farm Profitable

Mollison, Bill
Permaculture: A Designers Manual
Permaculture One
Permaculture Two

Moore, Raymond & Dorothy
Home Grown Kids

Morrison, Frank B.
Feeds and Feeding

Morrow, Rosemary
Earth User's Guide to Permaculture

Murphy, Bill
Greener Pastures on Your Side of the Fence

Nation, Allan
Grass Farmers
Pasture Profits with Stocker Cattle
Ranching for Profit

Olson, Michael
Metro Farm

Pollan, Stephen
Live Rich: Everything You Need to Know to Be Your Own Boss, Whomever You Work for
Die Broke : A Radical, Four-Part Financial Plan

Popcorn, Faith
Clicking: 17 Trends That Drive Your Business - and Your Life

Robinson, Jo
Why Grassfed Is Best

Rodale (staff)
Complete Book of Composting
**Encyclopedia of Organic Gardening*

Salatin, Joel
**Pastured Poultry Profits*
**Salad Bar Beef*
**You Can Farm: The Entrepreneur's Guide to Start & Succeed in a Farming Enterprise*

Savory, Allan
**Holistic Resource Management*

Scott, Steven K.
**Simple Steps to Impossible Dreams*

Siegmund, Otto H., Editor
**The Merck Veterinary Manual*

Smith, Burt
**Intensive Grazing Management: Forage, Animals, Men, Profits*

Smith, Russell
**Tree Crops: A Permanent Agriculture*

Staten, Hi W.
**Grasses and Grassland Farming*

Thompson, W.R.
**The Pasture Book*

Turner, Newman
**Fertility Pastures and Cover Crops*

Voisin, Andre'
**Grass Productivity*

Walters, Charles, Jr.
**An ACRES USA Primer*
**Weeds, Control Without Poisons*

Whatley, Booker T.
**How to Make $100,000 Farming 25 Acres*

Whitefield, Patrick
**How to Make a Forest Garden*

Willis, Harold
**The Coming Revolution in Agriculture*

Yeomans, P.A.
**Water for Every Farm: Yeomans Keyline Plan*

Ziglar, Zig
**See You at the Top*
**Reaching the Top*

Index

A

B

C

D

E

F

G

H

I

J

K

L

M

N

O

P

R

S

T

U

V

W

Z